ACS SYMPOSIUM SERIES **660**

Spices

Flavor Chemistry and Antioxidant Properties

Sara J. Risch, EDITOR
Science By Design

Chi-Tang Ho, EDITOR
Rutgers, The State University of New Jersey

Developed from a symposium sponsored by
the Division of Agricultural and Food Chemistry

American Chemical Society, Washington, DC

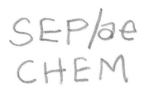

Library of Congress Cataloging-in-Publication Data

Spices: flavor chemistry and antioxidant properties / Sara J. Risch, editor,
Chi-Tang Ho, editor.

  p.   cm.—(ACS symposium series, ISSN 0097–6156; 660)

"Developed from a symposium sponsored by the Division of Agricultural
and Food Chemistry at the 211th National Meeting of the American Chemical
Society, New Orleans, Louisiana, March 24–28, 1996."

  ISBN 0–8412–3495–7

1. Spices—Congresses.   2. Condiments—Congresses.

  I. Risch, Sara J., 1958–   . II. Ho, Chi-Tang, 1944–   . III. American
Chemical Society. Division of Agricultural and Food Chemistry. IV. American
Chemical Society. Meeting (211th: 1996: New Orleans, La..) V. Series.

TP420.S65   1996
664′.5—dc21   96–52322
                                                                        CIP

This book is printed on acid-free, recycled paper.

PRINTED IN THE UNITED STATES OF AMERICA

Advisory Board

Foreword

THE ACS SYMPOSIUM SERIES was first published in 1974 to provide a mechanism for publishing symposia quickly in book form. The purpose of this series is to publish comprehensive books developed from symposia, which are usually "snapshots in time" of the current research being done on a topic, plus some review material on the topic. For this reason, it is necessary that the papers be published as quickly as possible.

Before a symposium-based book is put under contract, the proposed table of contents is reviewed for appropriateness to the topic and for comprehensiveness of the collection. Some papers are excluded at this point, and others are added to round out the scope of the volume. In addition, a draft of each paper is peer-reviewed prior to final acceptance or rejection. This anonymous review process is supervised by the organizer(s) of the symposium, who become the editor(s) of the book. The authors then revise their papers according to the recommendations of both the reviewers and the editors, prepare camera-ready copy, and submit the final papers to the editors, who check that all necessary revisions have been made.

As a rule, only original research papers and original review papers are included in the volumes. Verbatim reproductions of previously published papers are not accepted.

ACS BOOKS DEPARTMENT

Contents

INDEXES

Preface

SPICES ARE WIDELY USED IN FOOD PRODUCTS to create the distinctive flavor and character that is representative of different cuisines. The symposium on the flavor and antioxidant properties of spices was organized to look at new developments in the area of spice chemistry.

The nature of the volatile components in different spices is important in understanding the flavors they impart to food products. Research is being conducted to determine the significant volatile and nonvolatile compounds that create the distinct flavor of various spices. Variations in spices can occur depending on the region in which they are grown and the climatic conditions, which can dramatically influence their composition. This information is important in formulating food products and maintaining their consistency over time. As spices vary in their flavor profile, adjustments may need to be made in the formulation of a product or spice blend to maintain the desired flavor. Computer models are being developed to evaluate spices and spice blends to determine their composition and other relevant variables.

Another major area of interest today is the antioxidant properties of spices. Although spices have long been used to help preserve food, it has not been known what components give the preservative effect. Research in this area has expanded beyond the use of spices as preservatives to the potential health benefits they confer as antioxidants in the body. There is considerable evidence that specific components in spices may provide these beneficial effects. Research is being carried out to determine the active components and to explain the mechanism of action. With increasing interest in the use of food products to help maintain health and prevent disease, spices may play a significant role.

The symposium on which this book is based was sponsored by the Division of Agricultural and Food Chemistry at the 211th National Meeting of the American Chemical Society, which took place in New Orleans, Louisiana, from March 24–28, 1996. This symposium provided a forum for researchers from all over the world to present information on possible specific roles of spices in disease prevention.

We thank all of the participants in the symposium as well as the additional authors who have contributed papers to this book. All of our colleagues who gave of their time to assist in the review of the manuscripts are recognized, and their work is sincerely appreciated. The American Spice Trade Association was helpful in providing general information about spices and their usage.

SARA J. RISCH
Science By Design
505 North Lake Shore Drive, #3209
Chicago, IL 60611

CHI-TANG HO
Department of Food Science
Cook College
Rutgers, The State University of New Jersey
New Brunswick, NJ 08903–0231

November 22, 1996

GENERAL OVERVIEW AND METHODS

Chapter 1

Spices

Sources, Processing, and Chemistry

Sara J. Risch

Science By Design, 505 North Lake Shore Drive, #3209, Chicago, IL 60611

Spices are used throughout the world to season food products and create the unique characteristic flavors of different cuisines. Spices are grown primarily in tropical countries although the United States has recently started growing a number of different spices to supply the domestic needs. The use of spices has increased significantly over the past few years, due in part to the high level of interest in different types of foods that use a wide variety of spices. Interest is also developing in the ability of spices to act as antioxidants in addition to seasoning a product

Spices have long been important for food products. It was found that small amounts of various plants could be used to enhance the flavor of a food and also served to help preserve that food. In some cases spices were even used to mask spoilage or off-flavors in products. This use continued for centuries without any real understanding of how the spices were being effective. People simply understood that spices helped create a more desirable taste in the foods that were being prepared. Different cuisines are noted for using specific types of spices to create their characteristic flavors.

People also viewed spices as being important because in early history the entire economy of many regions was based solely on spice trade. Spices were the major item of trade and the region that could control spice trade dominated as a world power. That situation has changed dramatically with spices now accounting for less than 0.1 percent of world trade. Other raw materials as well as processed foods account for a larger percentage of world-wide trade than spices. There are still several countries that rely heavily on the trade of a specific spice. In Tanzania, the production of clove accounts for a large percentage of that country's economy. Tanzania grows about 2/3 of the world requirement for cloves. About half Granada's revenue is derived from the sales of nutmeg and mace. If you also consider vanilla, which is not technically a spice but is a very widely used flavoring material that comes from a plant grown in tropical regions, it is the second largest product in the Malagasy Republic.

There are 36 different herbs and spices that are generally recognized and commonly used. White and black pepper account for the largest amount of spices used in terms of dollar value. The next highest value spices consumed are cloves, nutmeg, cardamom, cinnamon ginger, mace and allspice. It should be noted that accurate figures for the value of spice imports and exports are difficult to track. Countries typically do not report spices separately from other food products.

Most spices that are imported into the United States come in whole or unprocessed. There are requirements that they meet minimum standards for purity and cleanliness. The Food and Drug Administration performs inspections on the incoming spices mainly to confirm that they are free from undesirable filth and that they are safe for use in food. The American Spice Trade Association (*1*) and AOAC (*2*) have defined specific tests for spices. A complete list of typical tests for the quality of herbs and spices was developed a number of years ago by Heath (*3*). This provides a comprehensive listing of tests that can give information on the volatile oil content as well as other quality related factors in spices.

Spices and herbs come from a number of different parts of plants. This can include the bark, seeds, berries and leaves of a plant. In the United States, spices have a legal definition that is spelled out in the Code of Federal Regulations (*4*). This states that spices are "any aromatic vegetable substance in whole, broken or ground form, except those substances which have been traditionally regarded as foods, such as onions, garlic and celery; whose significant function in food is seasoning rather than nutritional; that is true to name; and from which no portion of any volatile oil or other flavoring principle has been removed". The form that the spice is used in often depends on the application in the food product. In some cases it is desirable to have large, visible pieces of spice while in others the spice may be finely ground to be more easily incorporated into the product. Some of the applications for whole spices include peppercorns in sausages, bay leaves in soups and various seeds that are used on or in breads and other baked products including caraway seeds, sesame seeds and poppy seeds.

For the vast majority of applications, the spice is processed before using in a food product. General processing of spices usually involves some means of particle size reduction. Processors use different techniques to achieve the desired particle size. The process of grinding spices breaks down some of the cell structure to make the volatile oils in the plant more readily released when used in a product than if the spice were left in the whole form. Care must be taken during the grinding process to insure that the desired volatile oil level is maintained in the spice. One process that has been investigated to help preserve the quality of the spice is cryogenic grinding.

In addition to grinding or comminution of the plant part, a means of bacterial reduction is also needed. The different methods employed to reduce the bacterial load on spices, including irradiation, steam treatment and ethylene oxide, will be covered in a subsequent chapter in this book. A comprehensive review of individual spices including their flavor characters and various applications is included in the Source Book of Flavors by Reineccius (5). This reference also includes information about processing of spices as well as the specific volatile oils and the procedures that are used to extract the oils from spices. The flavor of herbs and spices is derived from the volatile oils that are present in the plant. The only exception to this is the capsicums which give the flavor to peppers. The essential oil content of herbs is often less than 1%. The content of other spices ranges up to 18% essential oil in clove buds. Essential oils are recovered from herbs and spices by steam distillation. This recovers the volatile oils but not the non-volatile components which contribute to the pungency of some spices such as ginger and pepper. Oleoresins are obtained by solvent extraction which yields both volatile and non-volatile components.

While spices have long been used in food products, their usage continues to grow at a significant rate. A recent estimate by the American Spice Trade Association indicates that the usage in the U. S. alone could reach one billion pounds by the year 2000 (ASTA personal communication, 1996). Interest in spices started growing after World War II and saw a sharp rise beginning in the 1980's. This can likely be attributed to first, the exposure that many people had to other spices in the 1940's and, more recently, to increased interest in different ethnic cuisines. It should be noted that the consumption figures that are reported include dehydrated onion and garlic even though by legal definition they are not spices. These items may be used as spices in a food in the U. S. but must be listed separately on an ingredient statement. The average consumption today is about 50% greater than the average a decade ago with the annual per capita consumption of spices in the United States over three pounds (ASTA, personal communication, 1996). The spices that have shown the largest increase in use include sesame seeds, oregano and paprika. It is likely that the growth in use of sesame seeds parallels the growth in fast food restaurants using sesame seeds on sandwich buns.

The other trends in spice consumption in the U.S. follow the growth of certain food categories. One area is in hot foods where the American public is becoming much more adventuresome in the foods that they will eat. Consumption of red pepper was up 105% from 1988 to 1992. Our bland foods are becoming much spicier. One company recently introduced a fudge sauce for ice cream that contains black pepper, red pepper and cinnamon in addition to dark chocolate. At the 1996 Institute of Food Technologists convention, one demonstration showed the use of red pepper sauce in brownies. In contrast, another area of growth is in the mild herbs including basil and oregano. This growth follows the increase in consumption of pizza and a proliferation of commercially available spaghetti sauces. As the desire for new and unique foods continues, it is expected the use of spices will continue to grow.

Spices are commonly grown in tropical regions of the world. This can present challenges to companies that are trying to find a consistent source of spices both from a quality and supply perspective. Farmers in the U. S. have started growing a number of different spices to meet our growing demand. When dehydrated onion and garlic are included in the figures, the U.S. in 1993 was able to supply 38% of its needs

(ASTA, personal communication, 1996). The other countries that are important sources of spices include India, Indonesia, Mexico, Guatemala, China and Canada. Canada has gained prominence for its production of mustard, caraway and coriander seeds.

With the growth in the use of spices, there has been continued research into the active components of spices not only from a flavor standpoint but also from functional perspective to explore the antioxidant properties of spices. Different spices are known to contain hundreds of active compounds. One area of interest is to identify the specific compounds in a spice that make the most significant contribution to the flavor of that product. A technique that has been applied to flavor compounds to determine their relative importance to flavor of a product is gas chromatography coupled with effluent sniffing, referred to as GC-O (gas chromatography-olfactometry). Several variations on this technique, including aroma extract dilution analysis, can be used to determine the compounds that are most important to the aroma. This uses the human nose to determine characteristic and importance of individual compounds to compliment gas chromatography which can only give an indication of relative amounts of different compounds but cannot give any information on importance to the flavor profile. It is known that flavor compounds can have widely varying sensory thresholds so having only the analytical data does not tell the entire story regarding importance of the individual compounds. The two specific spices that were evaluated and are presented in later chapters were saffron (Cadwallader and Baek) and fenugreek (Blank et el). This data can be used in developing flavor systems that may be able to replace or supplement spices in food products.

There is ongoing development of new analytical techniques to evaluate spices to develop specific information about their composition. These techniques may not be applicable only to spices but are serving to expand our knowledge of spices and their active components. One method is supercritical extraction which is used to more effectively remove the volatile compounds from a spice and eliminate the interference of other plant materials. Once the compounds are isolated, one detector that is being employed is specific to nitrogen and can aid in determining the amounts of capsaicinoids which contribute the heat or pungency to different types of peppers.

Spice companies and food manufacturers are also concerned about maintaining the consistency of the spices and spice blends that they are using. It is know that a number of factors can influence the abundance of the active components in spices. The climatic conditions can have a significant impact on spice quality from one year to the next. One chapter later in this book will address the influence of different growing conditions on the volatile compounds that are present (Randle). The same species of plant that is grown in different regions of the country can have a different volatile profile from that grown in another region. Understanding these differences is important for manufacturers to be able to maintain consistent flavor of a product when the spice being used may be sourced from different areas. A variety of plant species have been investigated to develop fingerprints of spices from different growing regions. This information can be used to determine authenticity of a spice and determine optimum blends that should be used to maintain the desired flavor profile. Statistical programs are also being developed to be used to determine the most likely constituents in a spice blend and aid in replication of a specific blend. This is presented in the chapter by Chen et al.

An area of considerable interest and the focus of many research projects is the antioxidant properties of spices. As was mentioned earlier, one of the earliest uses of spices was to help preserve foods. Little was understood at that time about the mode of action of the spices. It was simply known that they would help maintain the quality of the food to be stored. The area of antioxidant properties of spices is important for preserving the quality of foods but may also provide beneficial health effects for people consuming the spices. This area will be covered in much more detail in an overview of the antioxidant properties of spices that is presented later in this book in the chapter by Madsen et al. One spice component in particular that has received attention is curcumin. Several chapters address the effectiveness of curcumin and possible ways to synthesize new curcuminoids. The investigators have looked at the various ways in which the compounds may be effective as an antioxidant in different systems. This is an area that will continue to receive attention in the research community as people continue to look for the specific health benefits that certain foods may offer. Spices will remain important for the flavor that they impart to foods and may also gain significance for other benefits they have to offer.

Literature Cited

1. Analytical Methods of the American Spice Trade Association, 3rd ed. ASTA, Englewood Cliffs, NJ.,1978.
2. Official Methods of Analysis, 14th ed. AOAC, Washington, D.C. 1984.
3. Heath, H. B. *Can. Inst. Food Tech. J.* **1968**, *1(1)*, pp. 9 - 36.
4. Code of Federal Regulations, Title 21, U.S. Government Printing Office. 1995.
5. *Source Book of Flavors*; Reineccius, G. A., Ed.; Second ed.; Chapman and Hall, One Penn Plaza, New York, NY, 1994, pp 234 - 255.

Chapter 2

Methods of Bacterial Reduction in Spices

W. Leistritz

SpiceTec Limited, 185 Alexandra Way, Carol Stream, IL 60188

There are three major methods for bacterial reduction currently being used in the spice industry. These are ethylene oxide, irradiation and steam. Irradiation has received a great deal of press and although the popularity of this method is increasing, there is still whether or not the consumer will accept it. The industry has scientific theory but the consumer activists are concerned about the unknown future consequences of the technology. Another method that is not new bu has not been used in the U.S. is steam sterilization using superheated steam. This is a consumer friendly method that may be the solution that the world is looking for. The method is effective and viewed as being safe. A comparison of the three methods will be presented and the advantages and disadvantages of each method detailed.

Spices are grown all over the world, in many cases it is in areas where the cleanliness is not closely controlled. There is no effort made to limit contamination of the spices wither during growing or harvesting. While there are some specifications and regulations concerning imported spices, it is often necessary to process the spices to reduce the bacterial load prior to use in food products. The bacterial load varies with the type of spice and total plate counts in excess of one million colony forming units per gram have been reported in some spices (1). Different types of pepper are known to often have very high bacterial counts.

The original need sterilized spices came from a demand by the U.S. armed forces. Prior to World War II, they were seeking foods that could be held for long periods of time without spoiling. One source of the bacterial contamination that reduced the shelf-life of some products was the spice or seasoning being used. By reducing the level of bacteria in the spice, products could be made with significantly longer shelf-lives. The method for doing this needed to destroy the bacteria without changing the flavor or color of the spice. Sterilization techniques were developed and were so successful for products made for the armed forces, that it soon became a generally accepted technique for most spices and seasonings. All spices and seasonings do not require bacterial reduction, but there is a risk factor with all of the materials. The food manufacturer needs to weigh the risks of using materials that potentially have a high bacterial load with any negative perception there may be with the processing of the spice.

The treatment of spices to reduce the bacterial load is often referred to as sterilization. In some cases, the spice may actually be sterilized, however, a more accurate term is bacterial reduction. The first method that was developed to reduce the bacterial load on spices involved the use of ethylene oxide gas. One company that is using a version of this method today is Griffith MicroScience. The treatment consists of a timed, low temperature cycle of vacuum gassing with ethylene oxide or propylene oxide. The general process flow is described below.

1. Product is placed in a sealed chamber about the width of a standard U.S. pallet (40 x 42 in). The chamber may vary in length and height, but it is critical that it not be too wide.
2. A vacuum is pulled within the chamber.
3. The chamber is heated to bring the temperature up to 110 - 120 F.
4. Humidity is introduced. (It should be noted that the combination of heat and humidity are essential to activate the microorganisms within the spice.)
5. Ethylene oxide gas is introduced into the chamber.
6. The spice is held in the chamber under these conditions for a specified period of time which is directly related to the bulk density of the spice and the level of bacterial reduction that is desired.
7. The gas is expelled from the chamber and it is flushed several times with air prior to returning it to atmospheric pressure.
8. The product is then removed to a quarantine area until it is proven that no residual gas can be detected.

This is a proven method for spices, particularly with whole spices and seeds. However, the use of gas for ground spices is controversial today. This method is capable of reducing the bacterial load by at least $1x10^7$. The process may be repeated if the initial result does not meet customer requirements. Using ethylene oxide gas does require a loner time period than irradiation or steam, with a typical process requiring anywhere from 12 to 18 hours. Productivity is influenced by the size and amount of chambers available. The cost per pound of spice treated is based upon volume treated, bulk density and length of time required to obtain the desired bacterial reduction.

Irradiation has become a popular method of reducing bacteria in spices and seasonings since its introduction in the early 1980's, according to the personnel at SteriGenics in Schaumberg, IL. Through the controlled release of Cobalt 60, spices pass through a beam of gamma energy that destroys bacteria. The amount of energy is called the radiation absorbed dose and is measured in rads (1 rad = 100 ergs/g). Most spices are treated at a level of 1.0 - 1.5 rads which is less than 50% of the rate allowed by the Food and Drug Administration (FDA). The gamma process requires control of only one parameter and that is time of exposure. SteriGenics literature notes that their process is lethal to spoilage and pathogenic microorganisms in spices. The following steps describe the SteriGenics method for use of irradiation to reduce bacteria in spices.

1. Product is loaded onto a carrier and travels along a conveyor through a series of doors and locks into the cell area containing cobalt 60.

2. The product passes around and then through two shelves of isotopes where the greatest percentage of bacterial reduction occurs. While in the cell, the product is bombarded with energy at lower levels as well.

3. The product returns through the series of doors and locks on a timed schedule and program to its origin. The entire process takes between 5 and 15 hours, depending upon the initial bacterial load and the reduction that is required.

This is a cost effective method that offers a number of benefits to the consumer. The main benefit is its overall effectiveness. There is no moisture added to the spice and there are no adverse effects on the volatile oils or nonvolatile components so the flavor profile is not negatively impacted. The irradiation will completely penetrate packaging materials so that they can be processed without the need of opening a package.

The major concern with irradiation centers upon the fear of consumer activist groups, both domestic and international, that not enough data has been gathered to determine the long term effects of eating food products which have been irradiated or which contain ingredients that have been irradiated. For this reason, the use of irradiation of spices and seasonings has been banned in several foreign countries. Some U.S. food manufacturers have also made it a policy not to use any ingredients which have been irradiated. SteriGenics, the leading company offering irradiated spices is currently addressing these concerns both to governmental agencies and to consumer groups.

The use of steam to reduce the bacterial count in spices has been gaining momentum due to the concerns with the safety of both gas and irradiation. Steam is considered to be a safe form of bacterial reduction throughout the world. There are two methods of steam bacteria reduction being offered in the U.S. today. One of these is a wet method and the other is a dry process.

There are several methods of wet steam systems for bacterial reduction available in the U.S. These include the MicroMaster process used by McCormick, Fuchs Micro control and the Kikkoman continuous steam process. These systems use either closed chambers or continuous feed systems. The key to all of these systems is the ability to penetrate the whole spices or seeds with wet steam at a temperature of 100 C or greater for a specified period of time. The bulk density is another critical factor which helps determine the flow rate, moisture pick-up and eventually the cost per pound.

A consistent concern with the wet steam process is the ability to properly dry the spice during with one pass or treatment after sterilization. It may become necessary to re-dry the product at a lower temperature to achieve the desired moisture levels. The overall costs for using wet steam match those of using gas.

The use of dry steam (superheated) is the latest steam technology being used for spices and seasonings. The equipment to produce spices by this method was recently installed by SpiceTec Ltd. in their plant in Carol Stream, IL. It is the first use of this method in the United States. The process flow for this system is described below.

1. Continuous precleaned product is dropped through an airlock into a closed chamber that contains a shaking bed.

2. The product is sterilized quickly at a temperature between 108 and 125 C. During this process, the steam is super-heated and turns into saturated steam or vapor. A continuous airflow system begins the cooling of the spices immediately.
3. The product drops through an airlock into a second closed chamber where the cooling process is continued with ambient air without the development of condensation. Dust is filtered out during the entire process.

This process, which is viewed as being safe also offers a number of benefits. Continuous flow of the product yields excellent productivity. All different forms of spices can be treated including whole and ground spices, herbs and ground seasoning blends. The efficiency of the process offers the potential for cost savings when compared to both the gas and wet steam process.

All three methods of sterilization or bacterial reduction offer both benefits and disadvantages. Irradiation is the most effective means of bacterial reduction, however, there are consumer concerns about the safety of the product. Ethylene oxide is a proven method that has been in use for years but has recently come under the scrutiny of FDA and the Environmental Protection Agency (EPA). The wet steam process is not quite as effective as irradiation and the cost is slightly higher, however, it is regarded as a safe method. The dry steam method is highly effective and the newest method available. It is more expensive than irradiation but less expensive a process than both ethylene oxide and wet steam. No one method is perfect. The selection of the method to be used is up to the individually processor and must meet the needs of the customer.

1. Reineccius, G., Ed., *Source Book of Flavors,* Chapman and Hall, New York, **1994**, p. 244.

FLAVOR CHEMISTRY

Chapter 3

The Principal Flavor Components of Fenugreek (*Trigonella foenum-graecum* L.)

Imre Blank, Jianming Lin, Stéphanie Devaud, René Fumeaux, and Laurent B. Fay

Nestec Limited, Nestlé Research Center, Vers-chez-les-Blanc, P.O. Box 44, CH–1000 Lausanne 26, Switzerland

3-Hydroxy-4,5-dimethyl-2(5*H*)-furanone (sotolone) was established as the character impact flavor compound of fenugreek on the basis of gas chromatography-olfactometry. Sotolone was found to occur predominantly in the (5*S*) enantiomeric form (95%) and to have a $\delta^{13}C_{PDB}$ value of -19.7‰. About 2-25 ppm sotolone were determined in fenugreek of different origins using the isotope dilution assay technique. Sotolone was generated in model systems by thermally induced oxidative deamination of 4-hydroxy-L-isoleucine (HIL) using different carbonyl compounds. Up to 24 mol% yields were obtained by boiling HIL and methylglyoxal as reactive α-dicarbonyl at pH 5 for 10 h. Strecker degradation of HIL was found to be a competitive reaction resulting in the formation of 3-hydroxy-2-methylbutanal. The lactone of HIL, 3-amino-4,5-dimethyl-3,4-dihydro-2(5*H*)-furanone, was found to be a better precursor of sotolone. It generated about 36 mol% sotolone in the presence of methylglyoxal.

Fenugreek is the dried seed of *Trigonella foenum-graecum* L. (Fabaceae). The plant is an annual herb widely cultivated in Mediterranean countries and Asia (*1*). The pods contain about 10-20 yellowish seeds with an appetizing pleasing aroma. Toasted and ground fenugreek seed is an essential ingredient of curry powders and is often mixed with breadstuffs. It is used as a seasoning in food products such as pickles, chutneys, vanilla extracts, artificial maple syrup, and others.

Several volatile constituents have been reported in fenugreek (*2*), mainly terpenes and fatty acids. However, no systematic work has yet been published on compounds that contribute to the characteristic aroma of fenugreek. 3-Hydroxy-4,5-dimethyl-2(5*H*)-furanone (sotolone) was suggested as an important volatile constituent of fenugreek due to its seasoning-like flavor note (*3, 4*).

Sotolone is a powerful flavor compound found in several foods and spices (*5*). It contributes to the burnt/sweet note of aged sake (*6*), cane sugar (*7*), and coffee (*8*); to the spicy/curry note of lovage (*9*) and condiments (*10*); as well as to the nutty/sweet flavor of botrytized wines (*11*) and flor-sherry wines (*12*). The flavoring potential of sotolone is due to its low threshold values, that of 0.02 ng/l air (*8*), 0.3 µg/l water (detection/nasal, *10*), and 0.036 µg/l water (detection/retronasal, *13*).

The structural similarity between sotolone and 4-hydroxy-L-isoleucine (HIL), the most abundant free amino acid in fenugreek seeds, was pointed out (*14*, *15*). It was postulated that this unusual amino acid could be the precursor of sotolone in fenugreek (*15*) (Figure 1). This hypothesis has recently been supported by the fact that only the (5*S*) enantiomer of sotolone was found in fenugreek (*16*). This is in good agreement with the stereochemistry of HIL isolated from fenugreek, i.e. (2*S*,3*R*,4*S*) (*17*). However, these authors failed to discuss possible formation pathways.

OH NH₂

5 4 3 2 1 OH

O

(*2S,3R,4S*)-HIL

→

OH

4 3

5 1 2

O O

(*5S*)-Sotolone

Figure 1. Stereochemistry of 4-hydroxy-L-isoleucine (HIL) and sotolone found in the seeds of fenugreek (*Trigonella foenum-graecum* L.).

The purpose of the present study was to identify those volatile compounds which significantly contribute to the seasoning-like note of fenugreek using the approach of sensory directed chemical analysis. Gas chromatography in combination with olfactometry and mass spectrometry have been used as key steps of this approach (*18*, *19*). The formation of flavor impact compound(s) was studied in model systems using the quantification technique Isotope Dilution Assay (*20*, *21*). The mechanistic study was based on a hypothetical pathway proposed for the formation of sotolone via thermally induced oxidative deamination of HIL (*10*).

Experimental

Materials. Commercially available fenugreek seeds of different geographical origins and fenugreek oleoresin were used. Sotolone was from Aldrich and diethyl 2-methyl-3-oxobutanedioate, L-isoleucine, methylglyoxal (40% in water), phenylglyoxal, 2,3-butanedione, and 2,3-pentanedione from Fluka. Solvents and other chemicals were of analytical grade from Merck.

Synthesis. 3-Amino-4,5-dimethyl-3,4-dihydro-2(5*H*)-furanone hydrochloride (ADF) and 4-hydroxy-L-isoleucine (HIL) were obtained as diastereomeric mixtures by photochemical chlorination of L-isoleucine (Figure 2a), i.e. (3*S*,4*R*,5*R*/3*S*,4*R*,5*S*) and (2*S*,3*R*,4*R*/2*S*,3*R*,4*S*), respectively (*17*, *22*, *23*). [5,6-¹³C]-3-Hydroxy-4,5-dimethyl

2(5H)-furanone ([$^{13}C_2$]-sotolone) was prepared by condensation of diethyl 2-methyl-3-oxobutanedioate and [1,2-^{13}C]-acetaldehyde followed by lactonization and subsequent decarboxylation under strongly acidic conditions (Figure 2b) (10). The structures of the synthesized compounds were verified by elemental analysis, mass spectrometry (MS), and nuclear magnetic resonance spectroscopy measurements (^1H-NMR, ^{13}C-NMR) (24).

Gas Chromatography-Olfactometry. GC-O was performed on a Carlo Erba (Mega 2) equipped with a cold on-column injector, FID and a sniffing-port. Fused silica capillary columns of medium (DB-OV 1701) and high polarity (DB-FFAP) were used as previously described (9), both 30 m x 0.32 mm with a film thickness of 0.25 μm. The temperature program was: 50°C (2 min), 6°C/min to 180°C, 10°C/min to 240°C (10 min). Linear retention indices (RI) were calculated according to van den Dool and Kratz (25). The sensory significance of each odorant was evaluated and expressed as the flavour dilution (FD) factor (18).

Mass Spectrometry

Qualitative Analysis. Electron impact (EI) and positive chemical ionisation (PCI, ammonia) mass spectra were obtained on a Finnigan MAT 8430 mass spectrometer. MS-EI were generated at 70 eV and MS-CI at 150 eV with ammonia as the reagent gas. Non-volatile samples were directly introduced into the ion source held at 200 °C. Volatile components were introduced via a Hewlett-Packard HP-5890 gas chromatograph (GC-MS) using a cold on-column injection. DB-FFAP fused silica capillary columns were used (30 m x 0.25 mm, film thickness 0.25 μm). The carrier gas was helium (90 kPa). The temperature program was: 50°C (2 min), 4°C/min to 180°C, 10°C/min to 240°C (10 min).

Quantitative Analysis. Sotolone was quantified by isotope dilution assay using [$^{13}C_2$]-sotolone as internal standard (10, 24). Quantification experiments were performed with a HP-5971 GC-MS using the following conditions: DB-Wax capillary column (30 m x 0.25 mm, film thickness 0.25 μm); carrier gas: helium (100 kPa); splitless injection (250°C); temperature program: 20°C (0.5 min), 30°C/min to 100°C, 4°C/min to 145°C, 70°C/min to 220°C (10 min); EI ionisation at 70 eV; selected ion monitoring (SIM) of sotolone (m/z 128) and the labeled internal standard (m/z 130). The concentration of sotolone was calculated from the peak areas using a calibration factor of 1.1 (Figure 3). A good linearity was found in the concentration range 3-150 μg/ml. All samples were injected twice.

Fast Atom Bombardment (FAB-MS). This was applied to study the lactonization of HIL to ADF. FAB-MS was performed on a Finnigan MAT 8430 double focusing mass spectrometer. FAB ionisation was carried out with a saddle-field atom gun (Ion Tech, Teddington, UK) which was operated at 2 mA and 7-8 kV with xenon. Glycerol was used as matrix. The positive ions at m/z 130 (protonated molecular ion of ADF) and 148 (protonated molecular ion of HIL) were recorded.

Isotope Ratio Mass Spectrometry (GC-IRMS). This was performed with a Finnigan MAT delta S isotope MS coupled on-line with a Varian 3400 GC via a combustion interface. Isotope ratios were expressed as δ-values [‰] versus the PDP

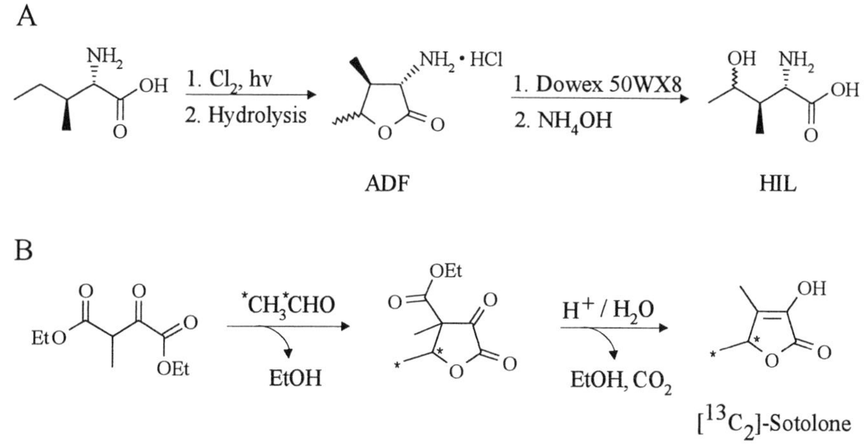

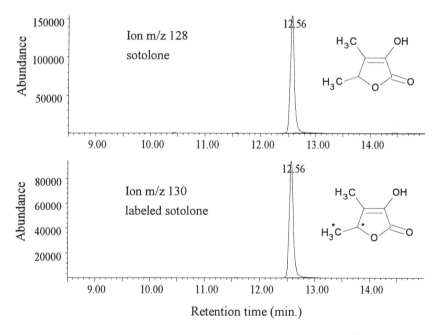

Figure 2. Synthesis of 4-hydroxy-L-isoleucine (HIL) and 3-amino-4,5-dimethyl-3,4-dihydro-2(5H)-furanone hydrochloride (ADF) (**A**), and [5,6-^{13}C]-3-hydroxy-4,5-dimethyl-2(5H)-furanone ([^{13}C$_2$]-sotolone, used as internal standard) (**B**).

Figure 3. Quantification of sotolone by isotope dilution assay using ^{13}C$_2$-sotolone as internal standard. The ions m/z 128 and m/z 130 were recorded by GC-MS operating in the selected ion monitoring mode. (Adapted from ref. 24.)

standard having a $[^{13}C]/[^{12}C]$ isotope ratio of 0.011237 for CO_2 yielded by combustion of fossil $CaCO_3$ (Peedee Belemnite). The GC was equipped with a DB-FFAP fused silica capillary column (30 m x 0.25 mm, film thickness 0.25 μm) using helium as carrier gas (100 kPa) and the split injection mode (220°C). The temperature program was the same as mentioned above.

Sample Preparation

Qualitative Analysis for GC-Olfactometry. The ground fenugreek seeds (100 g) were extracted with diethyl ether (Et_2O, 200 ml) by stirring the suspension for 5 h. The solvent was separated and the extraction was continued with additional solvent (200 ml) overnight. The extracts were combined, filtered and concentrated (80 ml) on a Vigreux column. Non-volatile by-products were separated by vacuum sublimation (VS), i.e. trapping volatiles under high vacuum conditions ($2 \cdot 10^{-4}$ mbar) into traps cooled with liquid N_2 (*21*). The sample was introduced drop-by-drop into the vacuum system to increase the yields. Then, Et_2O (30 ml) was added to the residue and the isolation procedure was repeated. The condensates of the traps were combined and concentrated to 1 ml on a Vigreux column.

Qualitative Analysis for GC-IRMS. Sotolone was isolated from fenugreek oleoresin. An aqueous solution of the oleoresin (35.5 g/100 ml) was extracted with Et_2O containing 5 % ethanol (5 x 100 ml). The organic phase was concentrated to 100 ml and extracted with Na_2CO_3 (0.5 mol/L, 3 x 100 ml). After acidification of the aqueous phase to pH 2 (HCl, 5 mol/L), sotolone was re-extracted with Et_2O (4 x 100 ml). The sample was dried over anhydrous Na_2SO_4 and concentrated to 20 ml. Sotolone was separated from nonvolatile by-products by sublimation in vacuum (10^{-2} mbar) (*21*). The condensates of the traps were collected and concentrated to 2 ml.

Qualitative Analysis for FAB-MS. HIL was dissolved in phosphate buffers (0.1 mol/L) with different pH values (pH 3.0, 4.0, 5.0, 6.0, and 7.0). The solutions were boiled for 1 h in sealed glass tubes. The samples were rapidly cooled down and directly analysed by FAB-MS.

Quantitative Analysis in Fenugreek Seeds (*10*). The material (5-10 g) was homogenized in water:ethanol (50 ml, 95:5). $[^{13}C_2]$-Sotolone (10-20 μg) was added as internal standard and the suspension was stirred for 30 min. After centrifugation (30 min, 10000 rpm), the supernatant was extracted with Et_2O. The acidic components were isolated with Na_2CO_3 (0.5 mol/L). The aqueous solution was acidified to pH 3 (5 mol/L HCl) and re-extracted with Et_2O. Finally, the organic layer was washed with saturated NaCl solution, dried over Na_2SO_4 and concentrated to about 0.2 ml using a Vigreux column and micro-distillation.

Quantitative Analysis in Model Experiments (*24*). HIL (2-10 mg) and ADF·HCl (2-10 mg) were each dissolved in a phosphate buffer (0.1 mol/L, pH 5.0). After adding the carbonyl reactant, the solution was boiled for 1 h in a sealed glass tube. The molar ratio of precursor to carbonyl was 1:10. Water and the internal standard ($[^{13}C_2]$-sotolone) were added to the cooled reaction mixture. The sample was saturated with NaCl and the pH adjusted to 4 (HCl, 1 mol/L). Sotolone was extracted from the reaction mixture with Et_2O for 8 h. The extract was dried (Na_2SO_4) and concentrated to 1 ml. All experiments were performed in duplicate.

Results and Discussion

Aroma Composition of Fenugreek. Liquid-liquid extraction using diethyl ether resulted in an aroma extract which represented the characteristic note of the original product, that i.e. is seasoning-like, spicy, herbaceous, and fenugreek-like. The representativeness of the samples before and after purification was checked by sensory evaluation.

GC-Olfactometry (GC-O) was used to detect the odor-active compounds present in the aroma extract of fenugreek. It is a simple but effective method to select those volatiles which contribute to the overall flavor (*18, 19*). On the basis of GC-O, seventeen odorants were detected in the original aroma extract (Figure 4). An aroma extract dilution analysis (AEDA) was applied to classify the aroma composition into three groups having different sensory relevance, "high" (no. 17), "medium" (nos. 4, 6, 16), and "background" (nos. 1-3, 5, 7-15). Hence, identification experiments were focused on the odorants belonging to the first two groups. Compound no. 17 was of particular interest because of its high FD-factor and characteristic seasoning-like note.

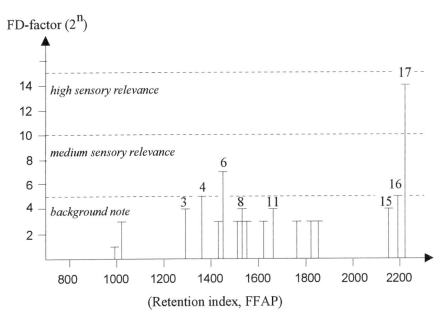

Figure 4. FD-chromatogram of an aroma extract obtained from fenugreek seeds.

The chemical structures (Figure 5) of the odorants were mainly elucidated by GC-MS (Table I). Compound no. 17 was identified as sotolone (Figure 6A). Sotolone is likely the character impact compound of fenugreek as indicated by the high FD-factor of 2^{14}, correspondingly low sensory threshold and characteristic aroma note.

Sotolone was detected by GC-O even after more than 10000-fold dilution of the original aroma extract. Its FD-factor was significantly higher than those of acetic acid (FD= 2^7), (Z)-1,5-octadiene-3-one (FD= 2^5), and 3-amino-4,5-dimethyl-3,4-dihydro-2(5H)-furanone (FD= 2^5) which belong to the group with medium sensory relevance (Figure 4).

The FD-factors of the remaining compounds were lower. They most likely contribute to the background of the fenugreek flavor. These odorants are short chain fatty acids (nos. 10, 11, 14), lipid degradation products (nos. 3, 12), and alkylated methoxypyrazines (nos. 5, 7). All odorants listed in Table I, except nos. 15 and 17, were identified for the first time as constituents of fenugreek aroma.

Table I. Odor-active Compounds Detected in an Aroma Extract of Fenugreek Seeds on the Basis of GC-O (n is the number of dilution steps)

No	Compound	Retention index		Aroma quality	FD-factor
		FFAP	OV-1701	(GC-O)	(2^n)
1	Diacetyl[a]	990	680	Buttery	1
2	Unknown	1020	n.d.	Fruity, metallic	3
3	1-Octene-3-one[b]	1296	1065	Mushroom-like	4
4	(Z)-1,5-Octadiene-3-one[b]	1369	1090	Metallic, geranium-like	5
5	3-Isopropyl-2- methoxypyrazine[b]	1428	1145	Roasty, earthy	3
6	Acetic acid[a]	1445	785	Acidic, pungent	7
7	3-Isobuty-2- methoxypyrazine[b]	1518	1235	Roasty, paprika-like	3
8	Linalool[a]	1543	1193	Flowery	4
9	Unknown	1554	n.d.	Sulfury, roasty	3
10	Butanoic acid[a]	1624	968	Sweaty, rancid	3
11	Isovaleric acid[a]	1663	1025	Sweaty, rancid	4
12	Unknown	1760	n.d.	Fatty	3
13	Unknown	1823	n.d.	Flowery, citrus-like	3
14	Caproic acid[a]	1845	1163	Musty	3
15	Eugenol[a]	2163	1500	Spicy	4
16	3-Amino-4,5-dimethyl-3,4-dihydro-2(5H)-furanone[a]	2190	n.d.	Seasoning-like	5
17	Sotolone[a]	2210	1350	Seasoning-like	14

[a] Identification by comparison with the reference compound on the basis of retention indices, aroma quality and GC-MS.

[b] Identification by comparison with the reference compound on the basis of retention indices and aroma quality. The amounts were too small for verification by GC-MS.

n.d. not determined

Terpenes and terpenoid compounds do not play a major role. Only linalool was detected by GC-O (no. 8). Most of the terpenes identified by GC-MS were odorless at the concentration present in the aroma extract, i.e. α- and β-pinene, sabinene, 3-carene, menthol, β-terpineol, cineol, anethol, β-terpinyl acetate, 1-*p*-

menthen-8-yl acetate, carvone, and several sesquiterpenes. Further volatiles of low or no sensory relevance were 1-pentanol, 1-hexanol, 2-methyl-2-butene-1-ol, 2-methyl-2-butenal, 2-pentylfuran, formic acid, propanoic acid, and further longer chain fatty acids, γ-butyrolactone and several 5-alkylated γ-lactones, 3-amino-4,5-dimethyl-2(3*H*)-furanone, and others.

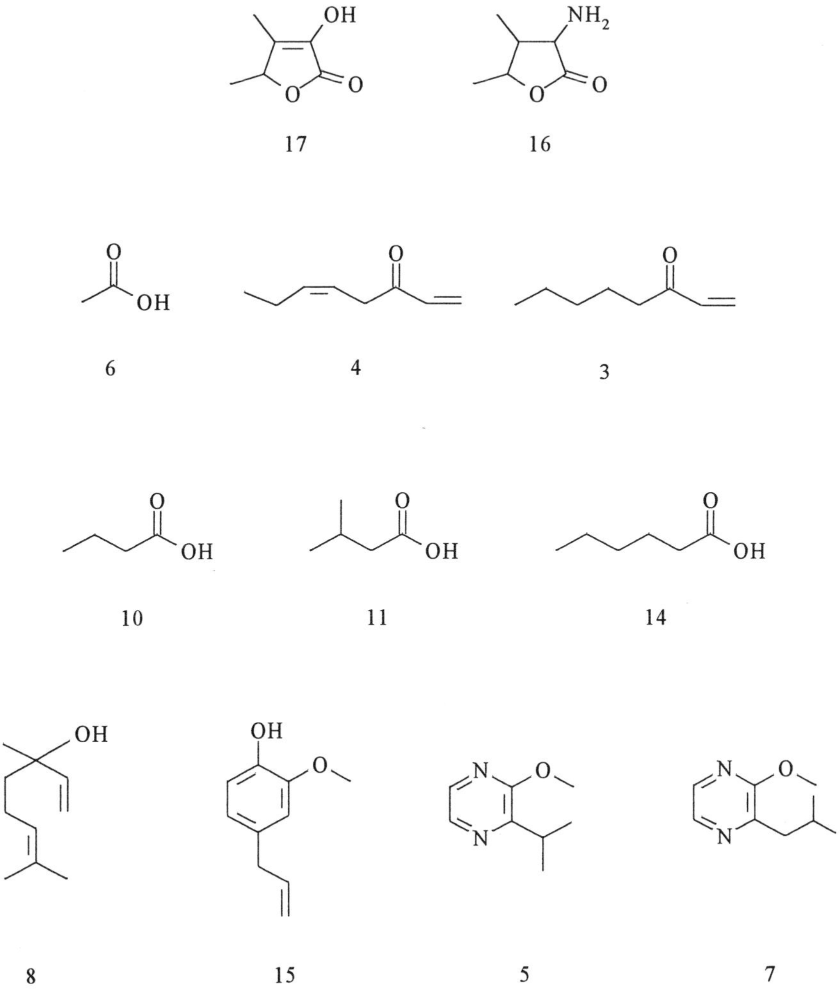

Figure 5. Sensory relevant compounds identified in the aroma extract of fenugreek. The numbers correspond to those in Table I.

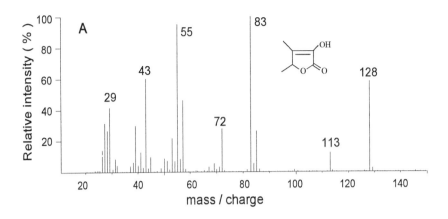

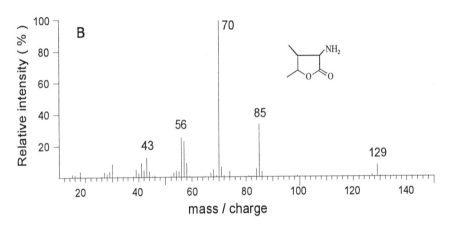

Figure 6. Mass spectra of 3-hydroxy-4,5-dimethyl-2(5*H*)-furanone (sotolone) (**A**) and of 3-amino-4,5-dimethyl-3,4-dihydro-2(5*H*)-furanone (**B**).

Stereoisomeric Characterisation of Sotolone. In fenugreek seeds, sotolone occurs predominantly in the (5*S*) enantiomeric form (95%). This is in good agreement with the data reported by Sauvaire et al. (*16*).

The δ^{13}C values of natural and synthesized sotolone were determined by isotope ratio mass spectrometry (IRMS) using the GC combustion technique (*26*). As shown in Table II, natural sotolone was characterised by a δ^{13}C$_{PDB}$ value of -19.7‰. On the contrary, a racemic mixture of synthesized sotolone showed a significantly higher value (-23.3‰).

Sotolone isolated from a commercial product revealed a δ^{13}C$_{PDB}$ value of -22.3‰ indicating a mixture of natural and synthetic compounds. This was confirmed by chirospecific GC analysis resulting in a ratio of 65:35 for (5*S*):(5*R*). Hence, both

GC-IRMS and chirospecific GC suggest that about 2/3 of the sotolone found in the commercially available liquid seasoning was contaminated with synthetic sotolone.

Table II. $\delta^{13}C$ Values (‰ PDP) of Sotolone of Synthetic and Natural Origin

Sotolone	$\delta^{13}C$ (‰)
Synthetic	-23.30 ± 0.20
Natural (fenugreek oleoresin)	-19.69 ± 0.20
Natural (fenugreek seeds)	-19.75 ± 0.20
Liquid seasoning (commercial)	-22.28 ± 0.20

Quantitation of Sotolone in Fenugreek. The concentration of sotolone was determined by isotope dilution assay (*10, 24*) using labelled sotolone as internal standard. As shown in Table III, the typical concentration range of sotolone was about 3-12 mg/kg fenugreek seeds. However, the amounts depend on the geographical origin. Fenugreek seeds from Egypt smelled more intensely and, in agreement with that, more sotolon was found in these samples. Some *Trigonella* species do not contain any sotolone as reported by Sauvaire et al. (*16*) and, consequently, they lack the characteristic seasoning-like note.

The sensory relevance of sotolone is due to its low threshold value of 0.3 µg/kg water (*10*). In fenugreek seeds, the concentration of sotolone is usually at least 3000 times higher than its threshold, thus indicating the sensory impact of sotolone to the overall flavor of fenugreek and products containing fenugreek. As shown in Table III, high amounts of sotolone were found in curry powder and some commercial liquid seasonings.

Table III. Concentration of Sotolone in Fenugreek and Products Containing Fenugreek

Sample	Sotolone [mg/kg]
Fenugreek, seed (Egypt, 1985)	25.1
Fenugreek, seed (Egypt, 1995)	12.2
Fenugreek, seed (Australia, 1991)	4.2
Fenugreek, seed (France, 1996)	3.3
Fenugreek, seed (India, 1996)	5.1
Fenugreek, seed (Turkey, 1995)	3.4
Curry powder (containing fenugreek)	39.7
Liquid seasoning A	1.4
Liquid seasoning B	88.9

Formation of Sotolone from Precursors

Precursors Present in the Aqueous Extracts of Fenugreek Seeds. The presence of glycosidically bound sotolone was examined by treating an aqueous extract of fenugreek with α- and β-glucosidases. As shown in Table IV, this enzymatic treatment did not significantly enhance the amounts of sotolone compared to the reference sample. On the other hand, boiling of the extract under acidic conditions (pH 2.4) for 1 h led to a more than 10 fold increase.

These trials indicate the presence of precursors in fenugreek that can be transformed to sotolone using specific conditions, particularly an acidic medium and heat treatment. 4-Hydroxy-L-isoleucine, known to be a characteristic amino acid of fenugreek (*14, 15*), may be one of these precursors.

Table IV. Formation of Sotolone from Precursors Present in the Aqueous Extracts of Fenugreek Seeds.

Addition to the aqueous extract	Reaction conditions			Sotolone [mg/kg]
	pH	T [°C]	t [min]	
None	6.8	25	120	18.9
α-Glucosidase (200 U)	6.8	25	120	19.2
β-Glucosidase (200 U)	6.8	25	120	20.2
None	2.4	100	60	252.3

SOURCE: Adapted from ref. 10.

Formation of Sotolone from 4-Hydroxy-L-isoleucine (HIL). The formation of sotolone was studied in phosphate-buffered model systems (pH 5.0) by reacting HIL with different mono- and α-dicarbonyl compounds at 100°C for 1 h. Both 2,3-butanedione and 2,3-pentanedione formed only low amounts of sotolone (Table V). Higher yields were achieved with the α-ketoaldehydes methylglyoxal (7.4 mol%) and phenylglyoxal (2.5 mol%), producing about 70-200 times more sotolone than the corresponding reaction with the α-diketones. Monocarbonyl compounds, such as propionaldehyde and phenylacetaldehyde, generated less than 0.1 mol% sotolone (*24, 27*).

Formation of Sotolone from 3-Amino-4,5-dimethyl-3,4-dihydro-2(5H)-furanone (ADF). The efficiency of ADF, the lactone of HIL, to generate sotolone was tested in the same model system as described above. As shown in Table V, significantly higher amounts of sotolone were generated from ADF as compared to HIL. Using methylglyoxal, the yields were increased from 64 μg (7.4 mol%) to 274 μg (35.9 mol%), thus indicating ADF to be a more efficient precursor than the free amino acid (HIL).

Table V. Formation of Sotolone from the Precursors 4-Hydroxy-L-isoleucine (HIL)[a] and 3-Amino-4,5-dimethyl-3,4-dihydro-2(5H)-furanone (ADF)[b] Present in Fenugreek Seeds.

α-Dicarbonyl	Precursor[c] HIL	Precursor[c] ADF	Sotolone[d] [μg/mg HIL]	Yield [mol%]
2,3-Butanedione	+	-	0.34 ± 0.03	< 0.1
2,3-Pentanedione	+	-	0.30 ± 0.03	< 0.1
Methylglyoxal[e]	+	-	64.2 ± 0.3	7.4
Phenylglyoxal	+	-	22.2 ± 0.3	2.5
2,3-Pentanedione	-	+	5.4 ± 0.3	0.7
Methylglyoxal	-	+	274.4 ± 3.4	35.9

[a] Control experiment (without α-carbonyl) yielded less than 0.01 mol% sotolone.

[b] Control experiment (without α-carbonyl) yielded 0.03 mol% sotolone.

[c] The molar ratio of precursor to α-dicarbonyl was 1:10.

[d] Data are means of at least two experiments, each injected twice.

[e] Control experiment (without HIL) yielded 0.07 μg sotolone.

Mechanism of the Formation of Sotolone (Figure 7, pathway A). The data reported above confirm the hypothesis of the formation of sotolone by thermally induced oxidative deamination of HIL (*10*). Acid catalyzed cyclization of HIL leads to the corresponding lactone (ADF) which reacts with an α-dicarbonyl (e.g. methylglyoxal) to form a Schiff base (pathway A). Rearrangement and subsequent hydrolysis gives rise to sotolone. The data show that α-dicarbonyls are capable of generating sotolone from both HIL and the lactone ADF. However, the latter is more efficient in producing sotolone. Furthermore, the relatively low yields achieved with HIL indicate an alternative degradation pathway.

Strecker Degradation of HIL as a Competitive Reaction (Figure 7, pathway B). The lower yields obtained with HIL might be explained by a partial Strecker degradation of the amino acid HIL in the presence of an active α-dicarbonyl, e.g. methylglyoxal. As shown in Figure 7, the amino-carbonyl reaction of HIL and methylglyoxal results in a Schiff base which may either cyclize or decompose via decarboxylation. The Strecker aldehyde of HIL, 3-hydroxy-2-methylbutanal, is released by hydrolysis and this compound was tentatively identified by GC-MS (*24*) as a mixture of diastereomers.

In the sample based on HIL and methylglyoxal, the ratio of Strecker aldehyde to sotolone was about 1:2 at pH 5 (Table VI). Consequently, the Strecker degradation of HIL is a competitive reaction to the formation of sotolone. In contrast, only traces of Strecker aldehyde were detected in the sample containing the lactone ADF, i.e. about 50 times less than in the reaction with HIL. The formation of sotolone from ADF is the favoured reaction, most likely due to the blocked carboxyl group.

Table VI. Ratio of Sotolone to 3-Hydroxy-2-methylbutanal Formed in Model Reactions Based on Methylglyoxal and the Precursors 4-Hydroxy-L-isoleucine (HIL) and 3-Amino-4,5-dimethyl-3,4-dihydro-2(5H)-furanone (ADF)

pH	HIL	ADF
3	1 : 17	1 : 50
5	1 : 2	1 : 50
6	1 : 2	1 : 30
7	1 : 1.5	1 : 40

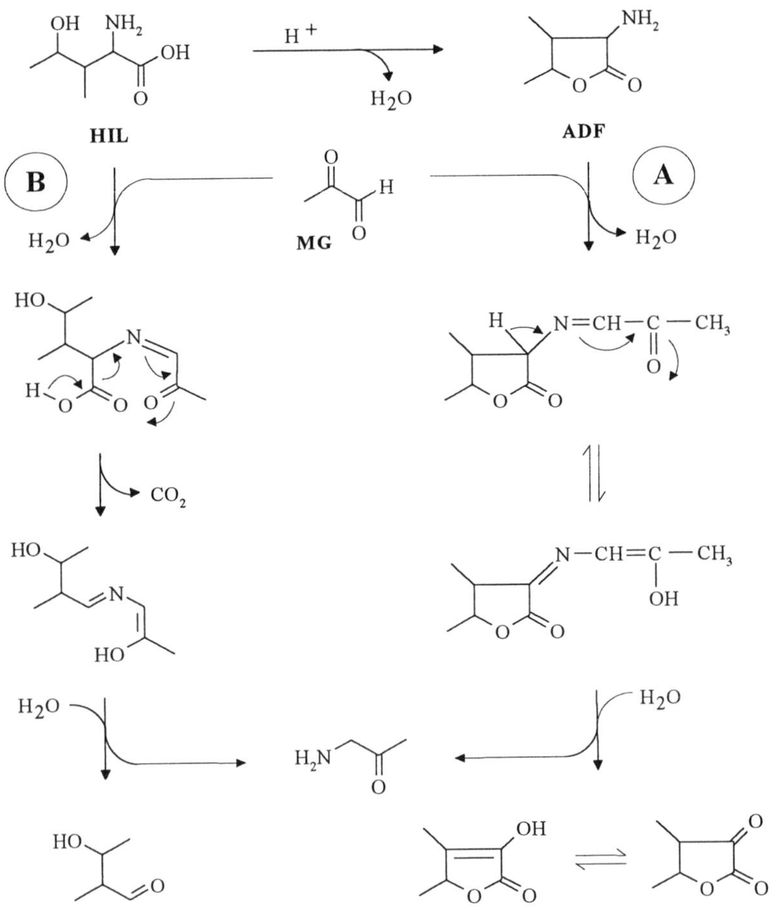

Figure 7. Formation of sotolone (pathway **A**) and 3-hydroxy-2-methylbutanal (pathway **B**) from 4-hydroxy-L-isoleucine (HIL) and 3-amino-4,5-dimethyl-3,4-dihydro-2(5H)-furanone (ADF) using methylglyoxal (MG) as carbonyl reactant.

Influence of the pH. It is well known that both lactonization and the formation of the Schiff base strongly depend on the pH of the reaction medium. As shown in Figure 8, the lactonization step (HIL → ADF) was favored under acidic conditions and the yields were about 50 % at pH 3, but only 10 % at pH 5. On the other hand, Schiff bases are readily formed under neutral and slightly basic conditions.

The reactivity of the carbonyl compound is another crucial parameter for the formation of the Schiff base. α-Ketoaldehydes are much more efficient than α-diketones (Table IV) and α-keto acids, which generate only low amounts of sotolone from HIL (*27*).

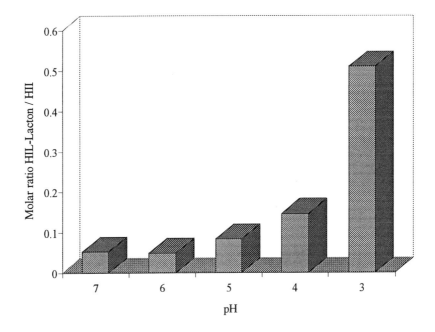

Figure 8. Lactonization of 4-hydroxy-L-isoleucine (HIL) as a function of pH.

To find the optimum pH, methylglyoxal was reacted with HIL and the lactone ADF, respectively. The reaction of methylglyoxal and HIL was favoured at pH 5 which apparently is the best compromise between the lactonization step and the reactivity of the amino group to form the Schiff base (Figure 7). Once the lactone (ADF) is formed, the amino-carbonyl reaction was favoured at pH 5-6. In general, the yields obtained with the lactone were significantly higher compared to the amino acid, particularly at pH 6 (40.2 mol%).

The reaction of HIL and methylglyoxal at pH 5 performed in water yielded 2.8 mol% sotolone compared to 7.4 mol% when using the phosphate buffered system. This suggests a catalytic effect of phosphate on the formation of sotolone from HIL.

Effect of Reaction Temperature and Time. The sotolone yields from HIL and methylglyoxal were strongly dependent on both reaction time and temperature. Significant amounts were generated above 70°C (Table VII). At a constant temperature of 100°C, the yield of sotolone continuously increased over a period of 10 h, with no significant increase thereafter (Table VIII). About 23 µg to 210 µg sotolone were generated from 30 min to 10 h which corresponds to 2.7 mol% and 23.8 mol%, respectively.

Methylglyoxal reacted with HIL at 50°C for 48 h resulted in 3.8 mol% sotolone, thus indicating that a long reaction time and mild reaction conditions are also suitable for generating significant amounts of sotolone. Therefore, hot climatic conditions might favour the formation of sotolone from HIL.

Table VII. Formation of Sotolone from HIL as a Function of the Reaction Temperature Using Methylglyoxal as Carbonyl Reactant [a]

Temperature [°C]	Sotolone[b] [µg/mg HIL]	Yield [mol%]
50	0.31 ± 0.01	0.03
60	1.15 ± 0.03	0.13
70	4.00 ± 0.05	0.46
80	12.5 ± 0.5	1.44
90	27.9 ± 1.8	3.2
100	64.2 ± 0.4	7.4

[a] Reaction conditions: phosphate buffer (0.1 mol/L, pH 5.0), 50-100°C, 1 h.
[b] Data are means of at least two experiments, each injected twice.

Table VIII. Formation of Sotolone from HIL as Affected by the Reaction Time Using Methylglyoxal as Carbonyl Reactant [a]

Time [h]	Sotolone[b] [µg/mg HIL]	Yield [mol%]
0.5	23.4 ± 0.8	2.7
1	64.2 ± 0.4	7.4
2	102.1 ± 4.5	11.7
5	170.2 ± 5.1	19.5
10	206.7 ± 10	23.8
15	208.3 ± 6.4	24.0
24	229.9 ± 5.3	26.4

[a] Reaction conditions: phosphate buffer (0.1 mol/L, pH 5.0), 100°C, 0.5-24 h.
[b] Data are means of at least two experiments, each injected twice.

Conclusion

The role of (5*S*)-sotolone as a character impact compound of fenugreek was corroborated and its formation from 4-hydroxy-L-isoleucine (HIL) via thermally induced oxidative deamination was substantiated. The lactone of HIL, 3-amino-4,5-dimethyl-3,4-dihydro-2(5*H*)-furanone (ADF), was found to be a better precursor than the amino acid. α-Ketoaldehydes were more effective in generating sotolone from both HIL and ADF than α-diketones. The reactivity of the dicarbonyl and the lactonization step are important parameters, particularly for the formation of the Schiff base. The transformation yields from HIL into sotolone greatly depend on the reaction conditions, such as temperature, time, pH and amount of dicarbonyl. High amounts of sotolone were obtained by boiling methylglyoxal with HIL for 10 h at pH 5 (24 mol%). Even better results (40 mol%) were achieved using ADF as precursor (1 h, pH 6), most likely due to inhibition of the Strecker degradation by the blocked carboxyl group.

Acknowledgments

We are grateful to Mr Y. Krebs for expert technical assistance, Dr. H. Schierbeek for performing the GC-IRMS measurements, and Dr. E. Prior for critically reading the manuscript. We also thank Prof. A. Mosandl, University of Frankfurt, Germany, for the chirospecific GC analysis.

Literature Cited

1. Lewis, Y.S. (Ed.), *Spices and Herbs for the Food Industry*; Food Trade Press: Orpington, England, **1984**; pp. 141-142.
2. Girardon, P.; Bessiere, J.M.; Baccou, J.C.; Sauvaire, Y. *Planta Med.*, **1985**, *51*, 533-534.
3. Rijkens, F.; Boelens, H. In *Proc. Int. Symp. Aroma Research*; Maarse, H.; Groenen, P.J., Eds.; Pudoc: Wageningen, The Netherlands, **1975**; pp. 203-220.
4. Girardon, P.; Sauvaire, Y.; Baccou, J.-C.; Bessiere, J.-M. *Lebensm. Wiss. Technol.* **1986**, *19*, 44-46.
5. Kobayashi, A. In *Flavor Chemistry. Trends and Developments*; Teranishi, R., Buttery, R.G., Shahidi, F., Eds.; ACS Symp. Ser. 388, American Chemical Society: Washington, DC, **1989**; pp. 49-59.
6. Takahashi, K.; Tadenuma, M.; Sato, S. *Agr. Biol. Chem.* **1976**, *40*, 325-330.
7. Tokitomo, Y.; Kobayashi, A.; Yamanishi, T.; Muraki, S. *Proc. Japan Acad.* **1980**, *56B*, 457-462.
8. Blank, I.; Sen, A.; Grosch W. *Z. Lebensm. Unters. Forsch.* **1992**, *195*, 239-245.
9. Blank, I.; Schieberle, P. *Flav. Fragr. J.* **1993**, *8*, 191-195.
10. Blank, I.; Schieberle, P.; Grosch W. In *Progress in Flavour and Precursor Studies*; Schreier, P.; Winterhalter, P., Eds.; Allur. Publ.: Wheaton, USA, **1993**; pp. 103-109.

11. Masuda, M.; Okawa, E.; Nishimura, K.; Yunome, H. *Agr. Biol. Chem.* **1984**, *48*, 2707-2710.
12. Martin, B.; Etievant, P.X.; Le Quere, J.L.; Schlich, P. *J. Agric. Food Chem.* **1992**, *40*, 475-478.
13. Wild, H. *Chem. Ind.*, **1988**, 580-586.
14. Fowden, L.; Pratt, H.M.; Smith, A. *Phytochemistry* **1973**, *12*, 1701-1711.
15. Sauvaire, Y.; Girardon, P.; Baccou, J.C.; Risterucci, A.M. *Phytochemistry* **1984**, *23*, 479-486.
16. Sauvaire, Y.; Brenac, P.; Guichard, E.; Fournier, N. In *Fourth International Workshop on Seeds, Volume 1*; Côme, D.; Corbineau, F., Eds.; ASFIS: Paris, France, **1993**; pp. 201-206.
17. Alcock, N.W.; Crout, D.H.; Gregorio, M.V.M.; Lee, E.; Pike, G.; Samuel, C.J. *Phytochemistry* **1989**, *28*, 1835-1841.
18. Grosch, W. *Trends Food Sci.Technol.* **1993**, *4*, 68-73.
19. Acree, T.E. In *Flavor Science. Sensible Principles and Techniques.* Acree, T.E.; Teranishi, R., Eds.; American Chemical Society: Washington, DC, **1993**; pp. 1-18.
20. Schieberle, P.; Grosch, W. *J. Agric. Food Chem.* **1987**, *35*, 252-257.
21. Sen, A.; Laskawy, G.; Schieberle, P.; Grosch, W. *J. Agric. Food Chem.* **1991**, *39*, 757-759.
22. Faulstich, H.; Dölling, J.; Michl, K.; Wieland T. *Liebigs Ann. Chem.* **1973**, 560-565
23. Hasan, M. In *New Trends in Natural Products Chemistry*. Rahman, A.; Le Quesne, P.W., Eds.; Studies in Organic Chemistry, Vol. *26*, Elsevier Sci. Publ.: Amsterdam, The Netherlands, **1986**; pp. 123-141.
24. Blank, I.; Lin, J.; Fumeaux, R, Welti, D.H.; Fay, L.B. *J. Agric. Food Chem.* **1996**, *44*, (in press).
25. van der Dool, H.; Kratz, P. *J. Chromatogr.* **1963**, *11*, 463-471.
26. Bruche, G.; Dietrich, A.; Mosandl, A. *Z. Lebensm. Unters. Forsch.* **1995**, *201*, 249-252.
27. Blank, I.; Lin, J.; Fay, L.B.; Fumeaux, R. In *Bioflavours 95. Analysis - Precursor Studies - Biotechnology.* Étiévant, X.; Schreier, P., Eds.; Les Colloques no. 75, INRA, Paris, **1995**; pp. 385-388.

Chapter 4

Vanilla

Daphna Havkin-Frenkel and Ruth Dorn

David Michael & Company, Inc., 10801 Decatur Road,
Philadelphia, PA 19154

Vanilla is the most widely used flavor in the food and confectionery industries. Worldwide annual consumption was 1900 tons in 1995, with 1400 tons imported to the USA alone.

Vanilla flavor is extracted from the cured bean of *Vanilla planifolia,* a member of the orchid family originated in Mexico. A small amount of beans is produced from yet another species - *Vanilla tahitensis.* Beans (pod-like fruit) are produced after 4-5 years of cultivation. Fruit maturation occurs after 10 months. The characteristic flavor and aroma develops in the fruit after a process called "curing", lasting an additional 3-6 months. We established tissue culture for vanilla, identified major components in the biosynthetic pathway of vanillin and found that tissue culture extract contains major components of vanilla flavor present in the bean. Hence, plant tissue culture may be an alternative method for vanilla flavor production, for studying the vanillin biosynthetic pathway and for micropropagation.

With a better understanding and ability to manipulate the biosynthetic pathway and to regulate cultivar quality and flowering, we may control production and consistent quality of the vanilla beans.

I. Agronomic Production Of Vanilla

Introduction. Vanilla is the world's most popular flavor. Its fruity, floral fragrance combined with a deep, aromatic body makes it unique and universally favored.

Vanilla is an epiphytic orchid native of the tropical region of Mexico. The flavoring material is obtained from dry cured pod-like fruits commercially called "beans". The generic name Vanilla is derived from the Spanish " vanillia", a diminutive of *vaina*, a pod. Its species name, *planifolia*, refers to the broad, flat leaf of the plant.

Vanilla became known in Europe following Cortez's conquest of the Aztec kingdom in 1519. Many centuries earlier, vanilla was a source of flavoring and used by the Aztec emperors who used vanilla to flavor a cocoa drink, which in the present day is hot chocolate. The Indians called vanilla "*Tlixochitl*", or black flower (6). The Spanish took vanilla back to their homeland and, shortly after, began manufacturing chocolate with vanilla flavoring. In England, Hugh Morgan recommended that vanilla flavoring should be used with chocolate to serve Queen Elizabeth I, following the example of the Aztecs.

For more than three centuries after Hernán Cortés lived, Mexico was the only vanilla producing country in the world. Many attempts were made to grow the plant in other tropical countries but these efforts failed since the plant or vine grew and flowered, but no fruits were produced. It was not until 1836 that Charles Morren, a Belgian botanist, discovered why vanilla was not able to produce fruit out of Mexico. The anatomy of the flower is such that self-pollination is impossible. Morren discovered that pollination is carried out by a tiny bee of the Mellipone family which lived in the vanilla growing region of Mexico. It is difficult for other insects to replace the tiny bees (3).

To achieve pollination one needs to remove the rostellum, a flower structure that is a modification of the stigma lying between the male and female organs and prevent access of the pollen to the stigma. Pollination is done by removing the rostellum with a sharp object, so that the pollen from the anther can be in contact with the stigma. Because the blossom lasts for a very short time (less than a day), pollination must take place as soon as the flower opens (6). Charles Morren was the first to propose hand-pollination and he was the first to produce vanilla beans outside of Mexico. This discovery laid the foundation for a new vanilla industry and broke Mexico's monopoly.

In 1841 Edmond Albium, a slave in the French-owned island of Reunion, discovered a rapid pollination method. With the use of a pointed tip of a small bamboo stick, he picked up the adhesive pollen and prying up the flap-like rostellum inside the flower, he pressed the male pollen mass onto the sticky female stigma. This method of pollination is still used commercially today. Using the technique described above, the French started vanilla cultivation on many of the islands in the Indian Ocean. Vanilla plantings were established in Reunion, Mauritius, Madagascar, Comoros, Thafila, Brazil, Jamaica and other islands in the West Indies. Vanilla was first introduced to the island of Java in 1819 by a Dutch scientist, using vines from Madagascar. Commercial production in central Java started in the early 20th century and is known as "Java Vanille" (3).

Today, the primary growing regions in the Indian Ocean are Madagascar, Comoros and Reunion. Beans from this area are called Bourbon and represent half of the world's vanilla production. Indonesia produces the other half of the world's supply with an increase in production and quality. Tahitian vanilla beans, which represent another variety, are grown mainly on the island of French Polynesia. Vanilla is also beginning to be grown on Tonga. Although vanilla started in Mexico, at present Mexico produces a small percent of the world's consumption (Fig. 1). Many other countries with suitable climates grow vanilla, but at present remain minor production regions (Fig.2).

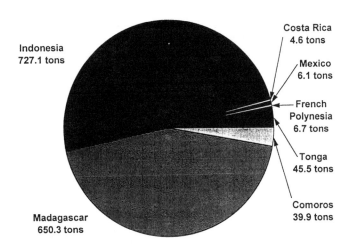

Figure 1. U.S General Imports of Vanilla Beans (U.S. Department of Commerce, 1995).

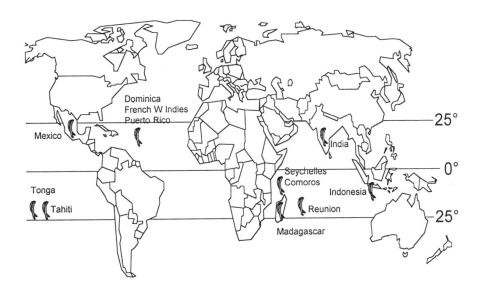

Figure 2. World Vanilla Production.

Cultivation

Species. Vanilla is a tropical climbing orchid. From more than 110 species that have been described, only 3 are of commercial value. They are V*anilla planifolia* Salisb Ames (Vanilla fragrans Andrews), *Vanilla tahitenesis*, and *Vanilla pompona*. Only *V. planifolia* and *Vanilla tahitenesis* are permitted to be used in food. V. *planifolia*, the most important commercially, is cultivated in all the vanilla growing areas, except for Tahiti and Hawaii. The beans are 10 to 25 cm long and 1 to 1.5 cm wide.

V. tahitenesis is indigenous to Tahiti and is different from *V. planifolia* by having slender stems and narrower leaves. The pods are also shorter than *V. planifolia* (6). The *V. tahitenesis* beans are perfumey and contain anisyl alcohol, anisyl aldehyde, anisic acid, and maybe heliotropine which are absent in *V. planifolia* beans. The beans of *V. tahitenesis* command higher market price than *V. planifolia* beans.

Economically, the least important is *Vanilla pompona*, also known as vanillon. It produces an inferior quality bean and in the past was used for perfume and tobacco flavor. The plant produces less beans than *V. planifolia* and has larger and wider leaves. However, *V. pompona* has important traits, including growth under more adverse conditions, and resistance to root rot disease that may be used in cross breeding.

Climate. Vanilla needs a warm and moist tropical climate with frequent, but not excessive rain. Under excessive rainfall, vanilla may be attacked by mildew and root disease. Under drought conditions, the vanilla plant can suffer considerable damage, which will result in a small number of flowers and low yield. In general, sloping land that has soil with high organic matter, high water holding capacity and access to irrigation in dry years will overcome the problems caused by weather (12).

Vanilla grows best in 40-50% of normal sunlight intensity. In excessive sunlight, the apical buds tend to lose moisture and the growth of the vines is stunted. Heavy shade will result in vines with thin stems, small beans, and reduced number of flowers and fruits. A shaded area, rainfall between 1000-3000 mm and an average temperature of 23° to 29°C year round are the preferred conditions.

Soil and Nutrients. Vanilla is a shallow rooting plant and grows well in well-drained humus-rich soil. A top layer of mulch helps keep the moisture in and the roots to spread. Soil parameters, such as texture and pH, appear to be more important than nutrients. In a heavy rainfall area, good drainage is essential. The best soil appears to be limestone with a pH of 6.0 to 7.0 and a deep layer of mulch, which provides the nutrients, over and around the roots. Although the level of calcium, nitrogen, potassium, phosphorous and micronutrients in vanilla are essential for the growth and production of beans, much more attention has been given to the type of mulch, because mulch appears to be more important than nutrients (11).

Support. The vanilla plant is a climbing orchid and needs mechanical support. Support is also needed to provide convenient access for pollination and harvesting. Fast growing trees with small branches are preferred and also provide shade. However, it is important that the supporting tree will not compete for water and nutrients. Trees native to the tropics are commonly used, such as avocado, coffee, annatto, orange, etc. In some cases, it is possible to use support trees which, themselves, can produce cash crops. In Madagascar, casuarina pine trees are also widely used. In most plantations, the support trees are planted before the vanilla plants (12). In greenhouses, vanilla can be grown falling down or hanging.

Propagation. Vanilla is commonly propagated from cuttings usually with 8-12 nodes. Small cuttings will produce vines, but the larger the cutting, the faster the plant will flower. The simplest way to plant is to lay the cutting on the ground and cover it with thick layers of mulch. However, this propagation method limits the amount of plant material that may be used. Also, the many origins of the cuttings represent genetic variability.

A vine becomes productive after 3-4 years, depending upon the weather and soil conditions, and the health of the vine. The vine flowers once a year for 5 or 6 successive years. After 8 to 10 years, productivity goes down. Drought, insufficient mulch, overexposure to sun, and over-pollination can contribute to a reduction in the productive period of the vine (4). When planning a new plantation, one has to account for the life expectancy of the vines. An hectare of land can hold about 4000 productive vines. The average yield is 1.5 to 2 kg green beans per vine. A mature good quality bean weighs between 15 to 30 grams (12).

Flowering, Fruit Set, Growth and Maturation. It is well known that vanilla vines need to reach a certain maturity before starting to flower. Apparently this is to accumulate growth factors or nutrients needed for blossoming. The factors affecting the timing and abundance of flowers are not fully understood. However, according to Childers et. al. (4) drought, temperature, pruning and the size of the original cutting may all influence the flowering. Removal of 4 to 6 inches of the apical bud 6 to 8 weeks before blossoming promotes flowering.

Vines usually blossom for 1 to 2 months. The flowers open early in the morning and last for one day. Pollination must be done the same day. Also, in a raceme (the flower cluster) only one flower opens per day and about 15 to 20 flowers open in daily succession, although not all of them will be fertile. It is desirable to pollinate to obtain between 5-10 beans because over-pollination will result in small beans and a short production life.

Because self-pollination is impossible to achieve (Fig.3), hand-pollination must be used to obtain a commercial crop. Hand-pollination is done as originally described by Edmond Albium in 1841 in Reunion, which consists of removing the rostellum back so that the pollen bearing anther can be pressed against the stigma. Once pollination takes place, the development of the pod ensuing with the enlargement of the ovary is completed in about 1 and a half to 2 months (3).

After the fruit has attained its full size, it takes an additional 5 to 6 months to mature and ripen. During the maturation period, the dry weight and the protein content increases. Many other physiological changes occur, as yet of unknown nature, that result in the production of flavor precursors.

Curing. The green, mature vanilla bean has no flavor or vanilla aroma. If the bean stays on the vine, the typical vanilla flavor will develop after a prolonged ripening period. However, in commercial practice, the characteristic flavor and aroma of vanilla beans is due to changes taking place during the curing process (2).

The curing process consists of four steps: 1. Killing 2. Sweating 3. Drying, and 4. Conditioning.

Killing designates a process aimed at abolishing tissue and cellular activity but retaining heat. Tolerant enzymatic activity is important for the curing process. The most common processes are hot water killing, sun killing, and oven killing. Beans are placed in a wire basket and dipped in 65°C water for a few minutes. Killing by heat can be done also in an oven or by exposure to the sun for a prolonged time. There is a patent for using freezing as another killing method (1). The killing process apparently entails the breaking of cell membranes and mixing of previously compartmentalized substrates and enzymes.

Killing is followed by "sweating", which allows excess moisture to escape. This reduces bacterial and fungal spoilage, and yet leaves enough moisture for enzymatic activity. This is the most important step. Glucosides, like glucovanillin, are broken down to vanillin. This is the step by which the typical characteristics of the vanilla bean flavor, aroma and color are developed. The duration of the sweating process is between 7-10 days.

Following the sweating step is "drying", which reduces the moisture content from 60-70% to 20-30%. The reduced moisture content is typical for a good quality cured bean and helps to eliminate any microbial activity and to secure a long shelf life. The amount of moisture also affects appearance, flexibility, and quality of the beans. In this step it is necessary not to obtain too hot a temperature, because some of the flavor and aroma components can evaporate or break down. The drying takes 7-10 days in the sun, 2-3 hours a day. Oven drying can be employed at 45°-50°C. Overdrying of the bean will result in an inferior quality bean.

The last stage is "conditioning", or aging. This can last from a month to several months. During this period, the vanilla bean obtains the typical "full flavor". Chemical reactions, such as oxidation and hydrolysis are the main events (12).

II.   Biotechnology Production

Recent developments in plant tissue culture and the need to gain control on bean quality and yield turns our attention to this method as another approach to solve some of the problems associated with the agronomic production of vanilla (5). Plant tissue culture can be used for three different objectives: 1. micropropagation 2. production of secondary products and 3. study of biosynthetic pathways.

Micropropagation. In most vanilla that is cultivated, commercial propagation is done by cutting. Several reports have shown the possibility of using tissue culture techniques to obtain clonal propagation and disease-free vanilla plants (8, 9). We used a different approach using seeds as starting material. We were able to germinate vanilla seeds under controlled sterile conditions in a petri dish. The seed prefers to germinate on the bean flesh (Fig. 4). Young green beans, 1 to 2 months old after being pollinated, were surface sterilized with 15% bleach. The beans were cut into segments and placed on Gamborg's, B-5 basal medium (7) free of hormone with the addition of cefotaxin and vancomycin at concentrations of 100 mg per liter. The bean sections were subcultured to a clean plate every two weeks. Germination of seeds embedded in the bean sections started after 3-6 months. The emerging embryos were transferred to rafts (mesh size 25 microns, Sigma Chemical Company, St. Louis, MO), using the same medium without agar (Fig. 5). Each embryo was cut into a few small pieces at each subculturing. After a few months, plantlets were formed and moved to the greenhouse for hardening, using orchid mix. In an established lab, the number of plantlets that can be produced is unlimited. Seeds from young beans germinate faster than those from mature beans probably because they have not become dormant.

The technique just described gives us the tools to produce a vast number of disease-free plantlets that may be used to select for desirable traits, such as high yield, early flowering and superior flavor and aroma of beans.

Secondary Product. In the last few years, there has been an attempt by Escagenetics Corporation to commercialize vanilla flavor by using plant cell culture (8). According to the published data, they obtained a yield of nearly 100 mg vanillin per liter of culture, or roughly 14 mg vanillin per one gram of dry weight of cell mass. It is likely that a several fold increase in vanillin yield would be necessary for production to be economical. Another but alternative biotechnological approach for the production of vanilla is being investigated by Westcott et al (13). They use vanilla roots to convert ferulic acid to vanillin. The limiting step here is where one can obtain natural ferulic acid for the bioconversion. We used embryo culture, reasoning that differentiated tissue will be more conducive than cell suspension to produce secondary metabolites. We were able to grow the embryo in up to a 10 L bioreactor with a marine impeller at 100 rpm, because the embryos are sensitive to high shear. The culture produced a larger number of the flavor components that are in vanilla bean (Fig. 6 (a) and (b)). The profile of the flavor components in cultured tissues is different than in the bean. It is not expected that undifferentiated cultured tissues will produce the same profile and similar proportions of flavor components that are produced in the bean which is differentiated and a specialized organ. We, therefore, concentrated on the production of vanillin, which has the highest economic value. We found that embryo tissue produced up to 0.16% vanillin and that a vanilla bean produced between 1-3% vanillin on a dry weight basis. At present, it appears that tissue culture production is appreciably lower and may not be economical.

Figure 3. Cluster of Vanilla Flowers.

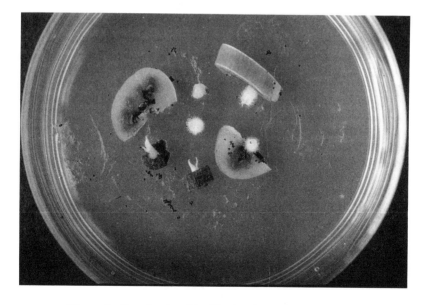

Figure 4. Germinating Vanilla Seeds on the Bean Flesh.

The germinating embryos were held on floating rafts to maximize access to air and nutrients. By the end of 60 days, the resulting plantlets were transferred to a pot with orchid mix in the greenhouse for hardening.

Figure 5. Growth of Germinating Embryos.

Vanillin Biosynthetic Pathway. Using the approach described previously, we grew the embryo culture to study the biosynthetic pathway of vanillin to find out the limiting step(s) in the production of the compound. After growing the culture and analyzing most of the extractable material, we propose that vanillin biosynthesis starts from chain shortening of p-hydroxycoumaric acid, resulting in the production of p-hydroxybenzoic acid, p-hydroxybenzyl aldehyde and p-hydroxybenzyl alcohol. We observed an accumulation of p-hydroxybenzyl alcohol up to 10% of the dry weight, suggesting that the enzyme that catalyzes the hydroxylation of p-hydroxybenzyl alcohol to 3, 4-dihydroxybenzyl alcohol is not active. In homogenized tissue, most of the p-hydroxybenzyl alcohol is converted to 3, 4-dihydroxybenzyl alcohol after a few hours, suggesting that in intact tissues the enzyme and the substrate are compartmentalized. Feeding experiments to the embryos with 3, 4-dihydroxybenzyl aldehyde, a by-product of 3,4-dihydroxybenzyl alcohol, led to the accumulation of vanillin, indicating that methylation is not a limiting step.

Given this information, understanding the limiting step can lay the foundation for future work. By isolating and purifying the enzyme that hydroxylates p-hydroxybenzyl alcohol and cloning the gene, we may obtain a culture with a high yield of vanillin that can be economically produced.

Future Opportunities For Biotechnolgy. We have shown that micropropagation of vanilla is possible with the use of tissue culture. This technology may be developed further to select for cultivars with superior traits. Importantly, it may be used to propagate transgenic tissues transfected, for example, with genes to increase the yield of vanillin, induce early flowering or disease resistance.

HPLC chromatogram of embryo culture extract.

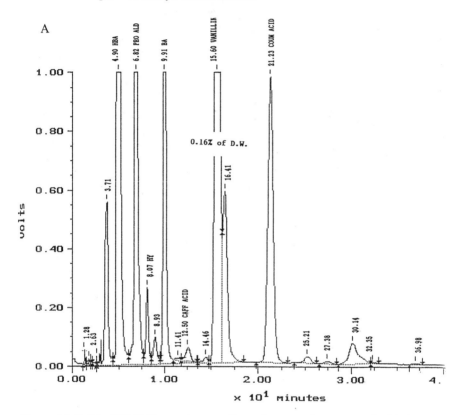

Figure 6A & 6B. HPLC Chromatogram. Isocratic separation was used with the mobile phase, 85% acidified water and 15% acidified methanol (1.25% acetic acid), controlled by a Waters 600E system. A Phenomenox Selectosil C18 column (250 x 4.6 mm) with 5 μm particle size was used. The peaks were detected at 280 nm using a Waters 994 photodiode array.

HPLC chromatogram of Bourbon vanilla beans extract (1 fold).

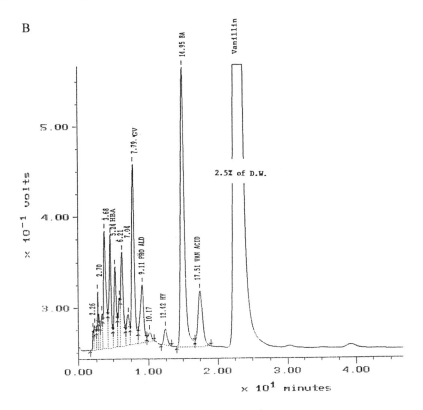

Figure 6. *Continued*

References

1. Ansaldi, Gwensele; Gil, Gerard; Le Petit, Jean, *U.S. Patent #4,956,192*, **Sep.11, 1990.**
2. Arana, F.E. *Circular No. 25, Washington, D. C., USDA* **1945**
3. Childers, N. F.; Cibes, H. R. *Circular No. 28, Washington, D. C., USDA* **1948**
4. Childers, N. F.; Cibes, H. R.; Hernandez-Medina, E. In *The orchids, a scientific survey*; Withner, C.L. Ed.; Ronald Press: New York, **1959**
5. Collin, H.A.; Watts, M. In *Handbook of plant cell culture - techniques for propagation and breeding, Vol. 1;* Evans, D. A.; Sharp, W. R.; Ammirato, P. V.; Yamada, Y. Eds; Macmillan Publishing Co.: New York, **1983**
6. Correll, D. S. *J Econ. Bot.* **1953**, 7, 291
7. Gamborg, O. L.; Miller, R. A.; Ojima, K. *Exp. Cell Res.* **1968**, 50, 148-151
8. Knuth, Mark E.; Sahai, Om P., *U.S. Patent #5,057,424*, **Oct. 15, 1991** and *U.S.Patent #5,068,184*, **Nov. 26, 1991.**
9. Kononowicz, H.; Janick, J. *Hortscience*, **1984**, 19, 58
10. Philip, V.J.; Nainar, S.A.Z. *J. Plant Physiol.* **1986**, 122, 211
11. Pureseglove, J.W.; Brown, E.G.; Green, C.L.; Robbins, S.R.J *Spices Vol 2* **1981**
12. Ranadive, A.S. In *Spices, herbs and edible fungi*; Charalambous, G. Ed.; Elsevier Science B.V.:Amsterdam, **1994**
13. Westcott, R.J.; Cheetham, P.S.J.; Barraclough, A.J. *Phytochem.* **1994**, 35(1), 135-138

Chapter 5

Onion Flavor Chemistry and Factors Influencing Flavor Intensity

William M. Randle

Department of Horticulture, 1111 Plant Science Building,
University of Georgia, Athens, GA 30602–7273

Although onions are an important vegetable and have nutritional value
in diets around the world, they are primarily consumed for their
distinctive flavor or their ability to enhance flavors in other foods.
Onion flavor is dominated by organosulfur compounds arising from the
enzymatic decomposition of S-alk(en)yl-L-cysteine S-oxide flavor
precursors following tissue disruption. Compounds arising from
precursor decomposition, such as the lachrymatory factor and various
thiosulfinates give onions their characteristic flavors. Sulfate is
absorbed by the plant and incorporated through cysteine to glutathione.
From glutathione, sulfur can proceed through several peptide pathways
and terminate in the synthesis of one of three flavor precursors. Flavor
intensity is governed by genetic factors within the onion and
environmental conditions under which the onions grow. Onion cultivars
differ in the ability to absorb sulfate and in the efficiency of
synthesizing flavor precursors. Increased sulfate fertility, higher
growing temperatures and dry growing conditions all contribute to
increased flavor intensity in onion.

Onions (*Allium cepa L.*) have world-wide importance, ranking second among all
vegetables in economic importance with an estimated value of $6 billion (*1*).
Although onions contribute significantly to the human diet and have therapeutic
properties, they are primarily consumed for their unique flavor or for their ability
to enhance the flavor of other foods. Onions are an ancient vegetable and can be
traced back through archeological records and early writings. Onions were used as
food, as medicines, in mummification, in art, and as spiritual objects (*2*).
Interestingly, onions have been domesticated for so long that they no longer have the

ability to exist in the wild and require human intervention for survival.   In contemporary society, onions weave their way through our diet and are consumed daily by most people. United States per capita consumption of onions in 1995 was approximately 18 pounds compared to 9 pounds in 1975 (per. com., National Onion Association).

Onion Flavor Chemistry

While compounds such as the water-soluble carbohydrates (sugars) and organic acids can contribute to the sensory experience when consuming onions, onion flavor is dominated by a special class of biologically active organosulfur compounds (3-4). Intact dry-bulb onions have little onion flavor or aroma. Flavor and aroma develops only when the onion is damaged or cut and flavor precursor compounds undergo enzymatic decomposition to form a variety of volatile sulfur compounds which give onions their characteristic taste and aroma. The first sulfur compound associated with onion flavor was identified in the 1890's (5). Pioneering research in this century among scientists such; as Chester Cavallito, Authur Stoll, Ewald Seebeck, nobel laureate Artturi Virtanen, Sigmund Schwimmer, Mendel Mazelis, George Freeman, Jane Lancaster, and Eric Block, has characterized many of the sulfur compounds contributing to flavor, the biosynthetic pathway for flavor precursor development, and the process by which flavor develops once the onion is cut or cooked.
    Onion flavor precursor formation begins with the uptake of sulfate (SO_4^{-2}) by the onion, its reduction to sulfide, and subsequent assimilation to cysteine by light-dependent reactions in the leaves of the plant (6). From cysteine, the sulfur can be further metabolized to form other sulfur-containing plant compounds. Sulfur's proposed entry into the flavor pathway is via glutathione. Early studies using radioactive isotopes suggested that sulfur passed through glutathione and was incorporated into S-2-carboxy propyl cysteine or S-2-carboxy propyl glutathione, eventually terminating into S-propenyl cysteine sulfoxide (7). Using radioactively labeled sulfate in pulse chase experiments, Lancaster and Shaw (8) demonstrated that sulfur was first incorporated into γ-glutamyl peptides as biosynthetic intermediates prior to terminating in the S-alk(en)yl cysteine sulfoxide precursors.
    There are 3 S-alk(en)yl cysteine sulfoxide flavor precursors in onions: S-(E)-1-propenyl cysteine sulfoxide is usually found in highest concentration and is responsible for tearing and pungency associated with onions; S-methyl cysteine sulfoxide which normally occurs in lesser concentrations; and S-propyl cysteine sulfoxide which is generally found is the lowest concentration (9). A fourth precursor, S-2-propenyl cysteine sulfoxide, is the predominate precursor of garlic, and found in other Alliums, but it is not detectable in onions. The pathways leading to the synthesis of each flavor precursor are not fully understood, nor do we understand the regulation of sulfur through these various pathways. Sulfur is thought to be transformed through several peptide intermediate pathways, unique to each of flavor precursors compounds (10; Fig. 1).
    The precursors are synthesized and stored in the cytoplasm of the plants's

cells (*11*).   Alliinase is compartmentalized in the cell's vacuole.   When the membrane of the vacuole is broken, alliinase is released and the precursors are broken down producing a chain of events.   Primary products include short-lived intermediate compounds, such as the 1-propenyl cysteine sulfoxide derived lachrymatory factor (LF, propanethial S-oxide), and other sulfenic acids from the different precursor species.   Other primary products are compounds pyruvate and ammonia which are more stable.   The LF, common to only 1-propenyl cysteine sulfoxide accumulating *Alliums*, produces the tearing, mouth burn, and pungency sensations (*12*).   A series of thiosulfinates are then formed which characterize the unique flavors and aromas of onion.   Early reports of di- and polysulfides and thiosulfonates were shown to be "artifacts" of hot injection port and gas chromatographic column (*13*).   The different flavor precursors give rise to different thiosulfinates which impart distinct flavors to the sensory experience (*14*).   For example, the propenyl/propyl thiosulfinates have green raw fresh onion flavors and the methyl/methyl thiosulfinates have a cabbage note (Table I).

The thiosulfinates then serve as the progenitor species of virtually all the sulfur compounds formed from the cut plant.   These compounds are unstable and undergo disassociation and rearrangement to form primary headspace volatiles (initial products formed from cut onions at room temperature such as the thiosulfinates), secondary volatiles (secondary products produced from the thiosulfinates at room temperature), and secondary solution components (products formed when thiosulfinates stand in solution at room temperatuare) (*15*).   Cut onions sitting on a kitchen counter or on a salad bar, therefore, change flavor over time.   For a thorough discussion of cut onion compounds, see Block, 1992.

Factors Influencing Flavor Intensity

Onion flavor intensity is governed by plant genetics and the environmental conditions in which the onion grows.   In the marketplace, onions vary widely in sulfur-based flavor intensity.

Cultivar Factors and Flavor Intensity.   Some onions are very pungent and aromatic while others can be eaten raw by the average person with relative ease.   Onion color has very little to do with how pungent it might be.   Red and yellow skinned onions can range from being mild to very pungent.   White onions, however, are usually only pungent to very pungent.   Some onions, such as the Grano and Granex cultivars can be grown mild with little flavor.   Other onions, such as the Danvers and Southport cultivars, are very pungent while some, such as the Sweet Spanish cultivars, fall between mild and pungent.   The genetic potential of a cultivar to absorb sulfur and synthesize the flavor precursors greatly determines how flavorful an onion will be.

While the heritability of flavor precursor accumulation in onions has not been determined, it is most likely a quantitatively inherited trait.   As described earlier, the biosynthetic pathway is complex with many peptide intermediates.   In addition, 11 proposed enzymes regulate compound synthesis in the flavor pathway (*17*).   Further empirical evidence to support quantitative inheritance of the flavor

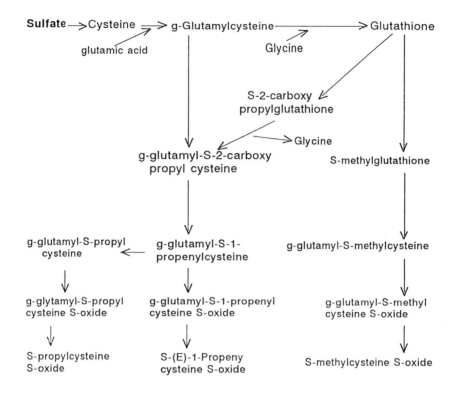

Figure 1. The proposed biosynthetic pathway for S-alk(en)yl cysteine sulfoxides in onions.

Table I. Sensory experience of different onion thiosulfinates and lachrymatory factor

Compound	Flavor Note
Methyl methane thiosulfinate	cabbage
Methyl propyl thiosulfinate	cabbage/onion
Propyl methyl thiosulfinate	cabbage/onion
Dipropyl thiosulfinate	green onion, chive
Propenyl methyl thiosulfinate	cabbage, onion, metallic
Methyl propenyl thiosulfinate	cabbage
Propenyl propyl thiosulfinate	raw fresh onion
Thiopropanal S-oxide	pungent, heat, mouth burn

precursors is the continuous variation found for the different flavor precursors within onion and the fact that the growing environment greatly influences precursor accumulation (18). Qualitatively inherited traits, which are governed by one to several genes, usually form discrete classes and are unaffected by the environment.

Cultivar differences can begin to be explained in several ways. Some cultivars accumulate 2 to 3 times the amount of sulfur as others, and sulfur accumulation can be a key element in determining onion flavor intensity (*19*). If we were to assume that cultivars assimilate the available sulfur into the flavor precursors with the same efficiency, differences among cultivars could be explained simply on how much sulfate enters the plant. However, cultivars differed in their efficiency at moving sulfur through the flavor pathway and in synthesizing the various precursors (*20*). The correlation, however, between total bulb sulfur accumulation and bulb pungency was poor among onions of broad genetic background, ranging between $r = -0.05$ and 0.36 (*21-22*). Some mild cultivars accumulate large amounts of sulfate, but do not accumulate corresponding concentrations of flavor precursors, while pungent cultivars will assimilate more of the available sulfur into flavor precursors (*23*). As an example, Savannah Sweet, a mild cultivar, had 0.40% total bulb sulfur (dry weight basis) and 4.05 mg/g fresh weight of total flavor precursors. On the other hand, Rio Grande, a pungent cultivar, had similar total bulb sulfur accumulation (0.43%), but accumulated 5.73 mg/g fresh weight total flavor precursors (a 42% increase in total precursor accumulation).

The ratio of the various flavor precursors also differs among cultivars. Flavor precursor composition and concentration can significantly influence flavor and flavor intensity. Some cultivars synthesize and accumulated mostly 1-propenyl cysteine sulfoxide, the lachrymator and pungency producing precursor, but little methyl or propyl cysteine sulfoxide (*24*). Other cultivars accumulate significant amount of methyl- and propyl- cysteine sulfoxide relative to the 1-propenyl precursor (*25*). The different precursor ratios give rise to different tastes and aroma. For example, if total precursor concentrations were equal, a cultivar with a ratio of 12:4:2 1-propenyl:methyl:propyl cysteine sulfoxide would be highly pungent and tear producing, while another cultivar with a 6:9:3 ratio would be less pungent, and have significant fresh onion flavor. Why cultivars differ in partitioning sulfur into the various pathways leading to the 3 cysteine sulfoxides or how environmental signals regulate synthesis among the pathways has yet to be determined.

Onion cultivars differ in alliinase concentration which may affect flavor development and intensity. Pukekohe Long Keeper, a pungent cultivar, had a 4 fold increase in bulb alliinase content compared to the mild Granex 33 cultivar (per. com., J.E. Lancaster, Crop and Food, NZ). In addition, onion alliinase decomposes the three cysteine sulfoxide flavor precursors at different rates (*26*). The kinetics of 1-Propenyl cysteine sulfoxide decomposition is quickest, followed by the methyl and propyl cysteine sulfoxides. *In vivo*, 1-propenyl cysteine sulfoxide decomposition was almost instantaneous and complete, whereas methyl cysteine sulfoxide and propyl cysteine sulfoxide were decomposed to a much lesser extent (*27*). Between 60 and 80% of the methyl cysteine sulfoxide, and approximately 50% of the propyl cysteine sulfoxide was left intact in onion macerate, even after 2 hours following maceration

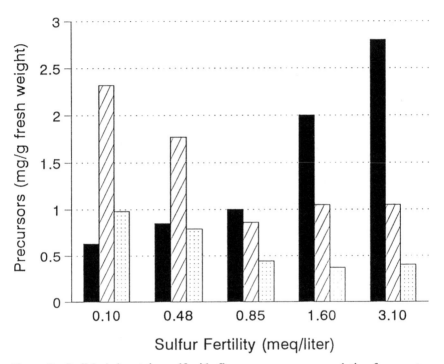

Figure 2. S-alk(en)yl cysteine sulfoxide flavor precursor accumulation from mature onion bulbs in response to increasing sulfur fertility from applied nutrient solutions. Solid bars are 1-propenyl cysteine sulfoxide. Lined bars are methyl cysteine sulfoxide. Dotted bars are propyl cysteine sulfoxide.

of bulb tissue. Cultivars also differed in the degree of decomposition of each precursor (28). Therefore, onion cultivars differing in flavor precursor composition will vary in flavor intensity do to their precursor concentration and degree to which the precursors are decomposed.

Flavor Distribution. Flavor is unequally distributed within the onion plant and is dependent on onion ontogeny. Within the bulb, there is a flavor gradient. The highest concentration of precursors occurs in the interior base of the bulb, while the lowest concentration of precursors are in the top outside scales (29). As the plant and bulb grows, flavor first increases in intensity, but then decreases as the bulb swells to maximum size and matures (30). Onion flavor and intensity also change during storage. Some cultivars increase in flavor intensity during storage while others decrease (31). The ratio of the different precursors can also change during storage. In one onion cultivar, the 1-propenyl thiosulfinates significantly increased in concentration during six months of storage while the methyl methane thiosulfinates completely disappeared (per com. Norman Schmidt, Dept. of Chem., Georgia Southern University).

Environmental Factors and Flavor Intensity. The environment in which an onion grows and develops will greatly influence how mild or pungent an individual cultivar will be. Yearly and locational flavor intensity differences among onion cultivars have been known and reported (32). When isolated individually, environmental factors such as sulfate availability, temperature, and water supply affect flavor intensity.

   Sulfate Fertility. Sulfate availability greatly influences onion flavor intensity (33). When onions were grown with sulfur fertility levels ranging from deficient to luxuriant, the concentration and ratio of the three flavor precursors changed (34; Fig 2). At sulfur fertility levels which cause sulfur deficiency symptoms in onion plants, methyl cysteine sulfoxide was the dominant precursor while 1-propenyl cysteine sulfoxide was a minor precursor. As sulfate fertility increased to adequate levels, 1-propenyl cysteine sulfoxide increased in concentration and methyl cysteine sulfoxide decreased in concentration. At luxuriant sulfate fertility, 1-propenyl cysteine sulfoxide was the dominant precursor while methyl cysteine sulfoxide became a minor precursor. Interestingly, total precursor concentration was similar from those onions grown at deficient sulfate fertility and those grown at luxuriant levels. It appeared that when onions were stressed for sulfate, the methyl cysteine sulfoxide biosynthetic pathway became a strong sink for the available sulfur and large amounts of methyl cysteine sulfoxide accumulated, even at the expense of plant growth and bulb development. The propanethial S-oxide and thiosulfinate concentrations followed the same pattern as the cysteine sulfoxide precursor concentrations when responding to increasing sulfate fertility (35). Sensory evaluation from professional taste panels confirmed the chemical analysis. As available sulfate levels increased in the growing environment, sensory notes such as

Table II. Aroma and flavor notes from 'Granex 33' onions grown at increasing sulfate fertility levels. Higher numbers represent more intense notes.

Attribute	low S	mod S	high S	excess S
Green aroma	6.4	6.0	5.9	6.6
Pungency aroma	7.4	8.0	7.2	8.4
Total aroma	8.3	8.9	7.7	9.5
Bitter	3.3	5.0	4.1	6.7
Green Flavor	6.8	6.9	6.2	7.3
Heat	4.9	7.2	6.6	8.1
Pungent flavor	5.7	8.1	7.2	9.4
Sweet	7.6	6.0	6.2	4.8
Total Flavor	8.2	9.5	9.2	11.1
Total Sulfur Flavor	5.2	7.8	7.0	8.2
Boiled Onion Flavor	4.4	3.1	3.7	3.1
Cabbage Flavor	3.4	3.1	3.1	3.1
Fresh Sulfur Flavor	7.0	7.5	6.7	7.5
Fruit Sulfur Flavor	4.7	4.6	4.6	4.7
Hydrogen Sulfide	1.4	1.8	1.5	2.2
Rubbery Sulfur	2.3	4.0	3.1	4.4

astotal aroma, bitterness, heat, pungency and total sulfur flavor also increased (Table II).

Growing Temperature. Temperature is important for onion growth and development. Minimum temperatures for bulbing are around 10°C and reach a maximum around 35°C. In an early study, although increasing temperature increased volatile sulfur compounds in onions, the author was unable to specify if flavor differences were due to temperature related plant growth (i.e., the faster the plant grew, the more pungent the plant became) or to the direct effect of temperature on flavor development (*36*). Recently, a study was completed where plants exposed to 4 different growing temperatures were: 1) grown for a specific length of time and harvested when the plants were of different developmental age, or, 2) grown to maturity which took increasing lengths of time as the temperature decreased. In both cases, increasing the growing temperature from 10°C to 31°C increased the pungency (as measured by enzymatically form pyruvate) of the onions, and the increase was linear in response to increasing temperature (Figure 3). The hotter the growing conditions, the more pungent the onions will be.

Water Supply. Growing onions under dry conditions will also increase bulb pungency compared to onions grown under well irrigated conditions. When onions were grown under natural rainfall or supplemented with irrigation water, the non-irrigated onions produced a higher volatile sulfur content compared to irrigated onions (*37*), or produced increased flavor strength as measured by volatile headspace analysis, enzymatically developed pyruvate, and sensory evaluation (*38*). As bulb size was smaller in the non-irrigated plots, it was thought that increased flavor strength was due to a concentration of the flavor precursor compounds in smaller cells. The exact mechanism for flavor increases in water-stressed plants, however, is yet to be determined.

Water usage and sulfate uptake by onions was also poorly correlated ($r = 0.09$; Randle, unpublished data). When plants were grown hydroponically to determine sulfate uptake requirements over the course of the growing season, water usage was greatly affected by daily differences in solar radiation while sulfate uptake was unaffected. The greater the solar radiation, the more water was transpired through the leaves.

Summary

The chemistry of onion flavor from sulfur compounds is quite complex. The biosynthetic pathway leading to the three forms of S-alk(en)yl-L-cysteine S-oxide flavor precursor synthesis is complicated and still being developed and proven. Onion cultivars differ in their ability to synthesize the flavor precursors and differ in the ratio of precursors synthesized. Each precursor gives rise to sulfenic acids and thiosulfinates which define different flavor experiences and flavor intensity. The environment in which the onions grows is also important in determining flavor intensity and composition. Increasing sulfate fertility, increasing the growing temperature, and/or decreasing the water supply will increase onion flavor strength.

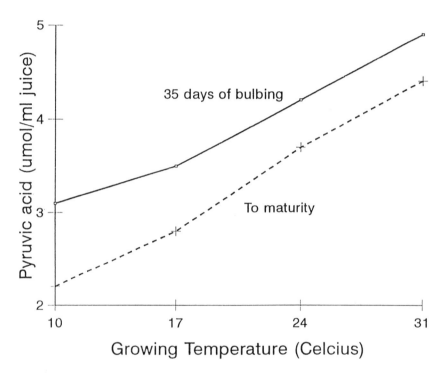

Figure 3.   Onion pungency (umol pyruvate per ml of onion juice) in response to increasing growing temperatures. Solid line represents plants grown at the different temperatures for 35 days in a bulbing photoperiod. Dashed line represents plants grown to maturity at the different temperatures.

Literature Cited

(*1*) FAO, *Production Yearbook*; Food and Agriculture Organization, Rome, 47, **1994**.

(*2*) Hanelt, P. In: *Onions and Allied Crops*; Brewster, J.L.; Rabinowitch, H.D., Ed. CRC Press, Inc. Boca Raton, FL, **1990**, Vol 1; .

(*3*) Block, E. *Angew. Chem. Int. Ed.* **1992**, *31*, 1135-1178.

(*4*) Darbyshire, B.; Steer, B.T. In: *Onions and Allied Crops*; Brewster, J.L.; Rabinowitch, H.D., Ed. CRC Press, Inc. Boca Raton, FL, **1990**, Vol 3; 1-16.

(*5*) Semmler, F.W. *Arch. Pharm.* **1892**, *230*, 434-443.

(*6*) Lancaster, J.E.; Boland, M.J. In: *Onions and Allied Crops*; Brewster, J.L.; Rabinowitch, H.D., Ed. CRC Press, Inc. Boca Raton, FL, **1990**, Vol 3; 33-72.

(*7*) Granroth, B. *Ann. Acad. Sci. Fenn. Ser.* **1970**, A2 *154*, 1-71.

(*8*) Lancaster, J.E.; Shaw, M.L. Phytochemistry. **1989**, 28, 455-460.

(*9*) Randle, W.M.; Lancaster, J.E.; Shaw, M.L.; Sutton, K.H.; Hay, R.L.; Bussard; M.L. *J. Amer. Soc. Hort. Sci.* **1995**, *120*, 1075-1081.

(*10*) Block, E. *Angew. Chem. Int. Ed.* **1992**, *31*, 1135-1178.

(*11*) Lancaster, J.E.; Collin, H.A. *Plant Sci. Lett.* **1981**, *22*, 169-176.

(*12*) Block, E.; Naganathan, S.; Putman, D.; Zhao, S. *J. Agrc. Food Chem.* **1992**, *40*, 2418-2430.

(*13*) Block, E. *Angew. Chem. Int. Ed.* **1992**, *31*, 1135-1178.

(*14*) Block, E.; Naganathan, S.; Putman, D.; Zhao, S. *J. Agrc. Food Chem.* **1992**, *40*, 2418-2430.

(*15*) Block, E. *Angew. Chem. Int. Ed.* **1992**, *31*, 1135-1178.

(*16*) Block, E. *Angew. Chem. Int. Ed.* **1992**, *31*, 1135-1178.

(*17*) Whitaker, J.R. *Adv. Food Res.* **1976**, *22*, 73-133.

(*18*) Randle, W.M.; Lancaster, J.E.; Shaw, M.L.; Sutton, K.H.; Hay, R.L.; Bussard; M.L. *J. Amer. Soc. Hort. Sci.* **1995**, *120*, 1075-1081.

(*19*) Randle, W.M. *Euphytica.* **1992**, *59*, 151-156.

(*21*) Randle, W.M. *Euphytica.* **1992**, *59*, 151-156.

(*22*) Randle, W.M.; Bussard, M.L. *J. Amer. Soc. Hort. Sci.* **1993**, *118*, 766-770.

(*23*) Randle, W.M.; Lancaster, J.E.; Shaw, M.L.; Sutton, K.H.; Hay, R.L.; Bussard; M.L. *J. Amer. Soc. Hort. Sci.* **1995**, *120*, 1075-1081.

(*24*) Thomas, D.J.; Parkin, K.L. *J. Agr. Food Chem.* **1994**, *42*, 1632-1638.

(*25*) Randle, W.M.; Lancaster, J.E.; Shaw, M.L.; Sutton, K.H.; Hay, R.L.; Bussard; M.L. *J. Amer. Soc. Hort. Sci.* **1995**, *120*, 1075-1081.

(*26*) Schwimmer, S.; Mazelis, M. *Arch. Biochem. and Biophy.* **1963**, *100*, 63-73.

(*27*) Lancaster, J.E.; Shaw, M.L.; Randle, W.M. *Proc. Natl. Onion Res. Conf.* **1995**, Madison, WI, 53-58.

(*28*) Lancaster, J.E.; Shaw, M.L.; Randle, W.M. *Proc. Natl. Onion Res. Conf.* **1995**, Madison, WI, 53-58.

(*29*) Freeman, G.G, *J. Sci. Food Agric.* **1975**, *26*, 471-481.

(*30*) Lancaster, J.E.; McCallion, B.B.; Shaw, M.L. *Physiol Plant.* **1986**, *66*, 293-297.

(*31*) Kopsell, D.E.; Randle, W.M. *HortScience.* **1995**, *31*, 766.

(*32*) Lancaster, J.E.; Reay, P.F.; Mann, J.D.; Bennett, W.D.; Sedcole, J.R. *New Zealand J. Exp. Agric.* **1988**, *16*, 279-285.
(*33*) Freeman, G.G.; Mossadeghi, N. *J. Sci. Food Agric.* **1970**, *21*, 610-615.
(*34*) Randle. W.M.; Block, E.; Littlejohn, M.; Putman, D.; Bussard, M.L. *J. Agric. Food Chem.* **1994**, *42*. 2085-2088.
(*35*) Randle, W.M.; Lancaster, J.E.; Shaw, M.L.; Sutton, K.H.; Hay, R.L.; Bussard; M.L. *J. Amer. Soc. Hort. Sci.* **1995**, *120*, 1075-1081.
(*36*) Platenius, H. *J. Agric. Res.* **1944**, *62*, 371-379.
(*37*) Platenius, H. *J. Agric. Res.* **1944**, *62*, 371-379.
(*38*) Freeman, G.G.; Mossadeghi, N. *J. Hort. Sci.* **1973**, *48*, 365-378.

Chapter 6

Contribution of Nonvolatile Sulfur-Containing Flavor Precursors of the Genus *Allium* to the Flavor of Thermally Processed *Allium* Vegetables

Tung-Hsi Yu

Department of Food Engineering, Da-Yeh Institute of Technology 112, Shan-jeau Road, Da-Tsuen, Chang-Hwa, Taiwan, Republic of China

This article discusses the contributions of nonvolatile sulfur-containing flavor precursors of the genus Allium to the flavor of thermally processed Allium vegetables through two approaches. In the first approach we have analyzed the volatile compounds generated from thermally processing blanched Allium vegetables. In the second approach we have analyzed the volatile compounds produced during thermal degradation or thermal interaction solutions of *S*-alk(en)ylcysteine sulfoxides, the major nonvolatile sulfur-containing flavor precursors of Allium vegetables. A large number of volatile compounds can be generated from thermally processed blanched Allium vegetables and from the thermal degradation or thermal interaction solutions of these sulfoxides, and many of these volatile compounds can be found in thermally processed Allium vegetables. We have demonstrated that the nonvolatile sulfur-containing flavor precursors of Allium vegetables can not only generate volatile compounds through thermal degradation or thermal interactions but also through the thermal degradation or interaction products of these sulfoxides to make important contributions to the flavor of thermally processed Allium vegetables.

Leaves or bulbs of the genus Allium, as represented by garlic (*Allium sativum* L.), onion (*Allium cepa* L.), shallot (*Allium ascalonicum* auct.), and welsh onion (*Allium fistulosum* L.), have been widely used at home and in the food industry as vegetables or flavoring materials because of their strong flavor properties. The characteristic aromas of the Allium species are mainly attributed to the sulfur-containing volatile compounds. Unlike the preformed volatile compounds, such as esters and terpene compounds in fruits and some spices which are biosynthesized as plants develop, the volatile components of the genus Allium are released from their nonvolatile sulfur-containing flavor precursors, especially *S*-alk(en)ylcysteine sulfoxides, by an enzymatic-mediated degradation that takes place when the plants are disrupted. Depending on the species, in the alk(en)yl groups the sulfoxides are mainly a combination of allyl, methyl, propyl, and 1-propenyl group (*1-3*).

Most of the research on the flavor of the Allium vegetables before 1991 focused mainly on the analysis of the volatile compounds of these vegetables, the formation of the volatile compounds in the genus Allium through the enzymic degradation and the transformation of the nonvolatile sulfur-containing flavor precursors, such as the *S*-alk(en)ylcysteine sulfoxides, and the stability of the thiosulfinates which are the enzymic reaction products of the *S*-alk(en)ylcysteine sulfoxides. There has been very little research on how amino acid characteristics affect flavor precursors.   Since *S*-alk(en)ylcysteine sulfoxides are amino acids, they should undergo thermal degradation or interactions with other food components, especially reducing sugars. The focus of this research has been on the contributions of nonvolatile flavor precursors of the genus Allium to the characteristic flavor of thermally processed Allium vegetables, especially garlic, shallot, and welsh onion. We approach this research in two ways. First, we used blanched Allium vegetables to see if significant amounts of volatile compounds could be generated from the nonvolatile sulfur-containing flavor precursors which were retained in the tissues of Allium vegetables after blanching these vegetables during thermal treatment, especially baking and frying. In our second approach, we synthesized four important *S*-alk(en)ylcysteine sulfoxides: *S*-allyl-, methyl-, propyl-, and 1-propenylcysteine sulfoxides.   These sulfoxides were then subjected to thermal degradation or interactions with other food components in the Allium vegetables to determine what their contributions were to the characteristic flavor of thermally processed Allium vegetables. The following reviews some of our research.

Volatile Compounds Generated from Thermally Treated Blanched Allium Vegetables

To determine the potential contribution of the nonvolatile sulfur-containing flavor precursors of the genus Allium to the flavor of thermally processed Allium vegetables, we used blanched Allium vegetables to which we applied thermal treatment. Blanching treatment can deactivate the flavor enzymes and retain most of the flavor precursors in the tissues of Allium vegetables. Since no significant amount of volatile compounds existed in the intact tissues of Allium vegetables, volatile compounds detected in thermally treated Allium vegetables could have been generated by thermally degrading the nonvolatile precursors in the tissues or through the thermal interactions of these precursors and other components, especially sugars, in the tissues.

We were able to identify some important volatile compounds in blanched and thermally treated blanched garlic slices, shallot slices and Welch onions as found in Tables I, II and III, respectively. As you can see in these three tables, no significant amounts of volatile compounds were identified from blanched garlic (BG), blanched shallot (BS), blanched green leaf of welsh onion (BGL), and blanched white sheath of welsh onion (BWS). These results prove that blanching treatment of Allium vegetables can deactivate the flavor enzymes in these vegetables efficiently and inhibit the enzymic formation of volatile compounds from the flavor precursors in Allium vegetables. However, you will note that in Tables I, II, and III, baking or frying treatment of the

Table I. Some important volatile compounds identified in garlic samples

Compound	BBG*	FBG*	BG*
	Yield, ppm		
Compounds Probably Generated from Thermal Degradation of Nonvolatile Flavor Precursors			
1-propene	7.12	10.55	2.20
acetaldehyde	1.85	25.24	0.70
methyl allyl sulfide	0.06	1.31	nd**
ally sulfide	1.16	8.99	0.09
methyl allyl disulfide	0.80	7.26	0.04
1,2-dithiacyclopent-3-ene	0.77	2.15	0.06
allyl disulfide	9.14	50.74	2.40
methyl allyl trisulfide	1.53	4.34	0.17
3-vinyl-4H-1,2-dithiin	0.49	1.44	0.14
2-vinyl-4H-1,3-dithiin	0.31	0.72	nd
allyl trisulfide	0.52	0.66	0.06
1,2,3,4-tetrathiepane	2.21	5.98	nd
dimethyl trithiepanes	2.42	7.63	0.03
Compounds Probably Generated from Thermal Interactions of Sugars and Nonvolatile Flavor Precursors			
2,5-dimethylpyrazine	0.51	0.69	nd
ethylmethylpyrazines	0.35	0.72	nd
3,5-diethyl-2-methylpyrazines	0.47	0.45	nd
Compounds Probably Generated from Thermal Interactions of Lipids and Nonvolatile Flavor Precursors			
4-heptenal	0.11	0.68	nd
2-ethylpyridine	0.12	0.20	nd
2-pentylfuran	nd	0.22	nd
methylethylpyridine	0.06	0.19	nd
phenylacetaldehyde	0.10	0.44	nd
buytlbenzene	0.06	0.19	nd
pentylbenzene	0.25	0.34	nd
hexylbenzene	0.43	0.62	nd
benzothiophene	0.70	0.40	nd

*   BBG: Baked Blanched Garlic; FBG: Fried Blanched Garlic;
    BG: Blanched Garlic. Data regenerated from ref. (*4*).
** nd: not detected

Table II.   Some important volatile compounds identified in shallot samples.

Compound	Yield, ppm		
	BBS*	FBS*	BS*
Compounds Probably Generated from Thermal Degradation of			
Nonvolatile Flavor Precursors			
1-propene	0.43	0.86	nd**
methanethiol	0.58	1.96	nd
dimethyl disulfide	0.46	1.91	nd
dipropyl disulfide	1.66	2.00	nd
1-propanethiol	0.29	0.30	0.03
3-methylthiophene	nd	0.78	nd
methyl propyl disulfide	2.49	8.50	0.01
dimethylthiophenes	4.83	15.44	nd
1-propenyl methyl disulfides	0.75	5.46	0.08
dimethyl trisulfide	3.52	7.28	nd
1-propenyl propyl disulfide	nd	1.42	nd
methyl propyl trisulfide	19.12	16.44	nd
dipropyl trisulfide	7.29	3.94	nd
Compounds Probably Generated from Thermal Interactions of Sugars			
and Nonvolatile Flavor precursors			
pyridine	nd	0.41	nd
2-pentylfuran	nd	5.80	nd
methylpyrazine	1.14	5.48	nd
dimethylpyrazines	13.64	59.24	nd
2,3-dimethylpyridine	1.23	nd	nd
ethyl methylpyrazines	6.11	15.38	nd
trimethylpyrazine	4.53	10.77	nd
5-ethyl-2-methylpyridine	nd	0.82	nd
2,6-diethylpyrazine	nd	3.38	nd
ethyl dimethylpyrazines	33.85	29.38	nd
methyl propylpyrazines	4.43	3.31	nd
dimethyl propylpyrazines	2.70	12.83	nd
2,3-dimethyl-5-[1-methylpropyl]pyrazine	1.51	nd	nd
3-ethyl-2,6-dimethylpyridine	0.94	nd	nd

*   BBS: Baked Blanched Shallot; FBS: Fried Blanched Shallot;
     BS: Blanched Shallot. Data regenerated from ref. (5).
** nd: not detected

Table III. Some important volatile compounds identified in welsh onion samples.

Compound	Yield, ppm					
	BBGL*	FBGL*	BGL*	BBWS*	FBWS*	BWS*
Compound Probably Generated from Thermal Degradation of						
Nonvolatile Flavor Precursors						
methanethiol	0.02	0.04	trace**	0.02	0.07	trace
1-propanethiol	0.08	nd***	0.02	0.08	nd	0.02
dimethyl disulfide	0.10	0.58	trace	0.05	0.67	nd
2-methylthiophene	0.02	0.01	nd	0.01	0.01	nd
methyl propyl disulfide	0.06	0.04	nd	0.05	0.03	nd
dimethylthiophenes	2.02	0.72	0.16	1.38	0.78	nd
1-propenyl methyl disulfides	0.63	2.64	nd	0.54	2.25	nd
dipropyl disulfide	0.02	0.03	nd	0.02	0.05	nd
dimethyl trisulfide	0.63	0.96	nd	0.72	1.44	nd
1-propenyl propyl disulfides	0.14	0.75	nd	0.71	1.01	nd
methyl propyl trisulfide	0.23	0.39	nd	0.33	0.69	nd
1-propenyl methyl trisulfide	0.08	0.21	nd	0.08	0.20	nd
dipropyl trisulfide	0.18	nd	nd	0.25	nd	nd
1-propenyl propyl trisulfide	0.08	0.21	trace	0.07	0.20	0.01
Compounds Probably Generated from Thermal Interactions of Sugars						
and Nonvolatile Flavor Precursors						
2-pentyl furan	nd	0.13	nd	nd	0.09	nd
dihydro-2-methyl-3-[2H]-furanon	0.24	nd	nd	0.28	nd	nd
dimethylpyrazines	0.04	nd	nd	0.05	nd	nd
3-(methylthio) propanal	nd	0.08	nd	nd	0.11	nd
2-ethyl-3,5-dimethylpyrazine	0.03	nd	nd	0.03	nd	nd
dimethylpyridine	0.04	nd	nd	0.01	nd	nd
2-acetylfuran	0.26	0.40	nd	0.34	0.49	nd
2-acetylpyrrole	0.08	nd	0.04	0.20	nd	0.01
2-methoxy-4-vinylphenol	0.27	0.07	0.04	0.03	0.02	0.05

*   BBGL: Baked Blanched Green Leaf; FBGL: Fried Blanched Green Leaf;
       BGL: Blanched Green Leaf; BBWS: Baked Blanched White Sheath;
       FBWS: Fried Blanched White Sheath; BWS: Blanched White Sheath.
       Data regenerated From ref. (5).

**   Trace: < 0.005 ppm

*** nd: not detected

blanched Allium vegetable slices generated significant amounts of volatile compounds. These volatile compounds are proposed to be mainly generated from the thermal degradation of nonvolatile flavor precursors in the Allium vegetables and the interactions of these precursors and other components, especially sugars and lipids, in these vegetables.

The major volatile compounds listed in Tables I, II and III separated into two groups; those that were probably generated from thermal degradation of nonvolatile flavor precursors and those generated from thermal interactions of nonvolatile flavor precursors of garlic and sugars. The major volatile compounds generated from thermally degraded nonvolatile flavor precursors of garlic listed in Table I were allyl disulfide, 1-propene, acetaldehyde, allyl sulfide, methyl allyl trisulfide, and other polysulfur-containing cyclic compounds. The major volatile compounds generated from thermal interactions of nonvolatile flavor precursors of garlic and sugars were pyrazines. Formation of pyrazines in these thermally treated blanched garlic slices are proposed through the Maillard type reactions of the amino-containing precursors and the reducing group-containing sugars existing in the garlic tissues. The major volatile compounds that were identified in baked blanched and fried blanched garlic slices which were probably generated from thermal interactions of nonvolatile flavor precursors of garlic and lipids were pyridines and alkyl benzenes.

Table II lists the major volatile compounds from baked blanched and fried blanched shallot slices generated from thermal degradation of nonvolatile flavor precursors of shallot were methyl propyl trisulfide, dimethylthiophenes, methyl propyl disulfide, and dipropyl trisulfide. The major volatile compounds that were probably generated from thermal interactions of nonvolatile flavor precursors of shallot and sugars were pyrazines, especially ethyl dimethyl pyrazines, dimethyl-pyrazines, ethyl methyl pyrazines, and trimethylpyrazine.

The major volatile compounds listed in Table III are from baked blanched and fried blanched welsh onion slices. The first group, generated from the thermal degradation of nonvolatile flavor precursors, were dimethylthiophenes, and alk(en)yl disulfides and trisulfides. The alk(en)yl group mentioned above could be methyl, propyl, and 1-propenyl. Those major volatile compounds generated from thermal interactions of nonvolatile flavor precursors of welsh onion and sugars were pyrazines, especially dimethylpyrazines, 2-methoxy-4-vinylphenol, 2-acetylfuran, and 2-acetylpyrrole.

Volatile Compounds Generated from Thermal Degradation or Thermal Interactions of of Alk(en)yl Cysteine Sulfoxides

To determine the potential contributions of the nonvolatile flavor precursors of the genus Allium to the flavor of thermally processed Allium vegetables, in our second approach we analyzed the volatile compounds generated from the thermal degradation of the *S*-alk(en)ylcysteine sulfoxides and from the interactions of these sulfoxides with other food components existing in Allium vegetables.   Four *S*-alk(en)ylcysteine sulfoxides were synthesized by us and were subjected to thermal degradation or thermal interactions; the volatile compounds generated were analyzed to determine what contribution these sulfoxides made to the flavor of thermally processed Allium vegetables. The sulfoxides

Table IV. Some important volatile compounds identified in the thermal degradation
or thermal interaction solutions of S-allylcysteine sulfoxide (alliin)

Compound	Yield, mg/mole of alliin		
	A*	A+G*	A+D*
1-propene	39.90	nd**	4.50
acetaldehyde	1199.60	16.50	nd
allyl alcohol	93.30	340.70	2193.60
ethyl acetate	232.30	nd	nd
acetic acid	72.70	4.30	nd
thiazole	9.90	13.30	42.50
acetal	26.40	nd	nd
dimethyl disulfide	11.10	nd	nd
2-methylthiazole	7.50	0.70	1.30
1,2-dithiacyclopentane	1.70	nd	15.90
methyl ethyl disulfide	62.00	nd	nd
methyl allyl disulfide	7.70	nd	nd
dimethyl trisulfide	11.50	nd	nd
formylthiophenes	3.50	29.90	6.00
2-acetylthiazole	359.50	140.30	16.20
methyl-1,2,3-trithiacyclopentane	188.00	nd	4.00
dimethyl tetrasulfide	5.30	nd	nd
ethyl methyltetrathiane	nd	1.10	24.90
methyl-1,2,3,4-tetrathiane	57.50	nd	nd
1,2,3,4-tetrathiepane	9.70	10.60	16.50
dimethyl tetrathianes	20.90	nd	nd
pyrazine	nd	4.10	nd
methylpyrazine	nd	43.80	nd
dimethylpyrazines	nd	10.00	nd
ethylpyrazine	nd	15.70	nd
ethyl methylpyrazines	nd	9.40	nd
2-acetylthiophene	nd	13.40	nd
ethyl dimethylpyrazines	nd	2.20	nd
methyl propylthiazoles	nd	30.60	nd
benzothiophene	nd	20.80	nd
phenylacetaldehyde	nd	1.90	nd
acetyl methylthiophenes	nd	6.00	nd
2-butylthiophene	nd	nd	31.50
2-pentylthiophene	nd	nd	6.10
2-pentylpyridine	nd	nd	78.40
2-hexylthiophene	nd	nd	61.10
2-hexanoylthiophene	nd	nd	31.50
2-pentylbenzaldehyde	nd	nd	394.10
2-formyl-2-pentylthiophene	nd	nd	110.20

* A: Alliin was thermally degraded in a pH 5 aqueous solution.
        Data regenerated from ref. (*6*).
   A+G: Alliin was thermally interacted with glucose in a pH 7.5 aqueous solution.
        Data regenerated from ref. (*7*).
   A+D: Alliin was thermally interacted with 2,4-decadienal in a pH 6 aqueous solution.
        Data regenerated from ref. (*8*).
** nd: not detected

Table V. Some important volatile compounds identified in the thermal
degradation or thermal interaction solutions of S-methylcysteine
sulfoxide

Compound	mg/mole of MeCySO	
	MeCySO[*]	MeCySO+G*
methanethiol	8.20	0.90
dimethyl disulfide	3320.80	2598.90
dimethyl trisulfide	21.60	59.80
thiazole	18.20	2.90
thiocyanic acid, methyl ester	65.30	1.30
methylthiol furan	157.50	86.60
pyrazine	69.80	114.60
methylpyrazine	8.90	264.00
ethylpyrazine	1.70	99.90
dimethylpyrazines	4.60	336.90
trimethylpyrazine	n.d.[**]	40.50
ethyl methylpyrazine	23.80	115.20
2,6-diethylpyrazine	19.30	6.30
ethenylpyrazine	5.40	53.20
ethyl dimethylpyrazines	6.50	74.90
ethenyl methylpyrazines	3.30	176.90
2-methylpyridine	4.30	3.10
2-acetylpyridine	n.d.	8.30
3-(methylthio)pyridine	17.40	5.70
1H-pyrrole	11.00	40.20

*   MeCySO : S-methyl-L-cysteine sulfoxide

     MeCySO+G : S-methyl-L-cysteine sulfoxide + glucose

**   nd: not detected

Table VI. Some important volatile compounds identified in the thermal
degradation or thermal interaction solutions of S-propylcysteine
sulfoxide

Compound	mg/mole of PrCySO	
	PrCySO*	PrCySO+G*
1-propanethiol	88.80	5.60
1,1'-thiobis propane	7.20	33.40
methyl propyl disulfide	8.60	38.70
dipropyl disulfide	6246.40	8367.80
dipropyl trisulfide	7088.80	3853.00
ethanethioic acid, S-propyl ester	44.00	2.70
pyrazine	4.50	60.90
methylpyrazine	nd**	126.80
dimethylpyrazines	3.60	119.40
ethylpyrazine	nd	70.90
2-ethyl-3-methylpyrazine	nd	13.30
2,6-diethylpyrazine	nd	9.20
ethenylpyrazine	4.60	21.10
ethyl dimethylpyrazines	5.50	65.30
methyl propylpyrazines	12.60	86.30
isopropenylpyrazine	nd	42.40
dimethyl propylpyrazine	nd	9.30
methyl-(1-propenyl)pyrazine	7.50	65.80
2-ethenyl-6-methylpyrazine	nd	87.50
3-(propylthio)pyridine	10.30	11.90
methyl (propylthio)pyridine	5.40	4.80
1H-pyrrole	nd	18.40

*   PrCySO : S-propyl-L-cysteine sulfoxide
PrCySO+G : S-propl-L-cysteine sulfoxide + glucose
**   nd: not detected

Table VII. Some important volatile compounds identified in the thermal
degradation or thermal interactions of S-(+)-cis-1-propenylcysteine
sulfoxide

Compound	mg/mole of PrenCySO	
	PrenCySO*	PrenCySO+ G*
(1-propenylthio) acetaldehyde	7.20	5.60
2-methyl-1,3-dithiane	15.80	3.00
methylthiophenes	33.60	253.00
2,4-dimethylthiophene	1.40	16.00
tetrahydrothiophene-3-one	nd**	4.40
thiophenecarboxaldehydes	2.40	58.00
5-methyl-2[5H]-thiophene	nd	9.80
formyl methylthiophenes	16.20	84.00
2-acetyl-5-methylthiophene	10.40	8.20
thiazole	3.20	29.40
4-methyl isothiazole	5.20	5.80
2,4-dimehtylthiazole	2.80	2.00
2-acetylthiazole	499.80	158.40
pyrazine	nd	52.60
methylpyrazine	5.00	86.40
2,6-dimethylpyrazine	4.60	51.40
ethyl methylpyrazines	nd	10.00
trimethylpyrazine	nd	5.20
3-ethyl-2,5-dimethylpyrazine	nd	10.20
3-methylpyridine	10.60	32.60
3,5-diethylpyridine	7.60	15.60
3-butylpyridine	nd	16.20

*   PrenCySO : (+)-S-cis-1-propenyl-L-cysteine sulfoxide
     PrenCySO+G : (+)-S-cis-1-propenyl-L-cysteine sulfoxide + glucose
**   nd: not detected

were *S*-allyl-, *S*-methyl-, *S*-propyl-, and (+)-*S-cis*-1-propenylcysteine sulfoxides.

Some important volatile compounds identified by thermal degradation or thermal interaction solutions of S-allylcysteine sulfoxide (alliin) are shown in Table IV. Thermal degradation of alliin mainly generated acetaldehyde, 2-acetylthiazole, ethyl acetate, methyl-1,2,3-trithiacyclopentane, and acetic acid. Thermal interaction of alliin and glucose mainly generated pyrazines, especially methylpyrazine, dimethylpyrazines, ethylpyrazine, ethyl methylpyrazines, methylpropylthiazoles, 2-acetylthiophene, and benzothiophene. Thermal interactions of alliin and 2,4-decadienal represented the major degradation products of vegetable oil containing linolenic acid, and were mainly alkylthiophenes, especially 2-formyl-5-pentylthiophene, 2-hexylthiophene, 2-butylthiophene, and 2-hexanoylthiophene, 2-pentylpyridine, and 2-pentyl-benzaldehyde.

In Table V we have identified some important volatile compounds from the thermal degradation or thermal interaction solutions of *S*-methylcysteine sulfoxide. Thermal degradation of *S*-methylcysteine sulfoxide mainly generated dimethyl disulfide and methylthio furan. Thermal interactions of *S*-methyl-cysteine sulfoxide and glucose mainly generated pyrazines, especially dimethylpyrazines, methylpyrazine, ethyl methylpyrazine, ethenyl methyl-pyrazines, pyrazine, and ethylpyrazine.

Table VI lists some important volatile compounds identified from the thermal degradation or thermal interaction solutions of *S*-propylcysteine sulfoxide. Thermal degradation of *S*-propylcysteine sulfoxide mainly generated dipropyl trisulfide and dipropyl disulfide. Thermal interactions of *S*-propylcysteine sulfoxide and glucose mainly generated pyrazines, especially methylpyrazine, dimethylpyrazines, methyl propylpyrazines, ethyl dimethyl-pyrazines, and 2-ethenyl-6-methylpyrazine.

Some important volatile compounds identified in the thermal degradation or thermal interaction solutions of *S-cis*-1-propenylcysteine sulfoxide are shown in Table VII. As shown in Table VII, thermal degradation of *S-cis*-1-propenylcysteine sulfoxide mainly generated 2-acetylthiazole and methyl-thiophenes. The addition of glucose to the aqueous solution of *S-cis*-1-propenylcysteine sulfoxide increased the yield of thiophenes but decreased the yield of 2-acetylthiozole. Thermal interactions of *S-cis*-1-propenylcysteine sulfoxide and glucose mainly generated pyrazines, especially methylpyrazine, pyrazine, 2,6-dimethyl pyrazine, ethyl methylpyrazine, and 3-ethyl-2,5-dimethyl-pyrazine, 3,5-diethylpyridine, and 3-butylpyridine.

We have demonstrated that a large number of volatile compounds generated from the thermal degradation of *S*-alk(en)ylcysteine sulfoxides or from thermal interactions of these sulfoxides with food components can also be found in the thermally processed Allium vegetables (*5,9*), proof that the nonvolatile sulfur-containing flavor precursors of Allium vegetables can generate volatile compounds through thermal degradation or thermal interactions but also that the thermal degradation or thermal interaction products of these sulfoxides make an important contribution to the flavor of thermally processed Allium vegetable.

Literature Cited

1. Block, E.; Naganathan, S.; Putman, D.; Ziao, S.-H. *J. Agric. Food Chem.* **1992**, 40, 2418.

2. Block, E.; Putman, D.; Zhao, S.-H. *J. Agric. Food Chem.* **1992**, 40, 2431.
3. Block, E.; Naganathan S.; Putman D.; Zhao S.-H. *Pure & Appl. Chem.* **1993**, 65, 625.
4. Yu, T.H.; Lin, L.Y.; Ho, C.-T. *J. Agric. Food Chem.* **1994**, 42, 1342.
5. Chen, Y.N. *Study on the contribution of flavor precursors to the flavor formation of thermally processed shallot and welsh onion,* **1996**, M.S. dissertation of Da-Yeh Institute of Technology, Taiwan, ROC.
6. Yu, T.H.; Wu, C.M.; Rosen, R.T.; Hartman, T.G.; Ho, C.-T. *J. Agric. Food Chem.* **1994**, 42, 146.
7. Yu, T.H.; Wu, C.M.; Ho, C.-T. *J. Agric. Food Chem.* **1994**, 42, 1005.
8. Yu, T.H.; Lee, M.H.; Wu, C.M.; Ho, C.-T. In *Lipids in Food   Flavors,* Ho.C.T.; Hartman, T.G. Ed.; ACS Symp. Ser. 558; American Chemical Society: Washington, D.C., **1994**, 61-76.
9. Yu, T.H.; Wu, C,M,; Ho, C.-T. *J. Agric. Food Chem.* **1993**, 800.

ANALYTICAL TECHNIQUES

Chapter 7

Characterization of Saffron Flavor by Aroma Extract Dilution Analysis

Keith R. Cadwallader, Hyung Hee Baek, and Min Cai

Department of Food Science and Technology, Mississippi Agricultural and Forestry Experiment Station, Mississippi State University, Herzer Building, Stone Boulevard, Mississippi State, MS 39762-9805

Volatile compounds were isolated from Spanish "Mancha Superior" saffron by simultaneous steam distillation-solvent extraction (SDE) and direct solvent extraction (DE). Extracts were analyzed by gas chromatography (GC)-mass spectrometry, GC-olfactometry (GC-O), and aroma extract dilution analysis (AEDA). Total ion chromatograms of SDE and DE extracts were different with respect to levels and types of volatiles present; however, both extracts had distinct saffron-like qualities. A total of 25 aroma-active components were consistently detected by GC-O and AEDA, with 18 common to both SDE and DE extracts. One compound tentatively identified as 2-hydroxy-4,4,6-trimethyl-2,5-cyclohexadien-1-one (saffron, dried hay-like) had the highest $\log_3$(flavor dilution factor) in both extracts, followed by 2,6,6-trimethyl-1,3-cyclohexadiene-1-carboxaldehyde (safranal) (saffron, tea-like) and an unknown compound having a saffron, dried hay-like aroma.

Saffron, the dried dark-red stigmas of *Crocus sativus* L. flowers, is used to impart both color and flavor to foods. Considered to be one of the most expensive spices, saffron is primarily produced in Spain (*1,2*), but is grown commercially in several other countries. Production, chemistry, adulteration, and quality assurance aspects of saffron have been reviewed (*2,3*). The intense yellow color of saffron primarily is due to the presence of crocin, a mixture of water soluble carotenoid glycosides (*4-8*). The primary taste-active component of saffron is picrocrocin, a bitter glycoside of 2,6,6-trimethyl-4-hydroxy-1-formyl-1-cyclohexene (*4-8*). Hydrolysis of picrocrocin leads to the formation of safranal, the major component of the essential oil of saffron (*9*). Safranal, having a typical saffron-like aroma, has been generally considered to be the character-impact component of saffron (*10*). Rödel and Petrzika (*11*), using gas chromatography-olfactometry (GC-O), confirmed the importance of safranal to saffron aroma; however, they detected several unidentified aroma-active components.

GC-O is used for the detection of aroma-active components in a volatile extract (*12*). The relative aroma intensity of each component can be determined by aroma extract dilution analysis (AEDA), which involves GC-O evaluation of a serial dilution series of a volatile extract (*13*). From these results, a flavor dilution (FD) factor, the highest dilution at which a specific aroma compound was last detected by GC-O, is obtained for each aroma-active component of the extract. FD factors are then used to arrange the aroma-active compounds according to their significance in the extract, thus providing a better understanding of the role each compound plays in the overall flavor of the food. A critical evaluation of AEDA can be found elsewhere (*14-15*). AEDA has been employed for the identification of important aroma components in wide variety of foods, e.g. lobster (*16*), crab (*17*), beef (*18*), and bread crust (*19*).

The present study deals with the identification of predominant aroma-active compounds in saffron by AEDA. In order to assure that the results of AEDA were representative of saffron aroma, volatile components were isolated by two extraction techniques.

Materials & Methods

Materials. Dried saffron, type "Mancha Superior", was purchased from two domestic spice suppliers. A total of three 1 oz. samples were obtained for this study. Original packing date of this material was 1995. Samples were stored in the dark at room temperature until analysis.

Standard flavor compounds were purchased from Aldrich Chemical Co. (Milwaukee, WI).

Simultaneous Steam Distillation-Solvent Extraction (SDE). Whole saffron stigmas (0.5 g), 50 mL of deodorized-distilled water, and 10 µL of internal standard solution (3.07 mg/mL 3-Heptanol in methanol) were continuously extracted with 10 mL of redistilled dichloromethane in a micro scale SDE apparatus (catalog no. 16051, Chrompack, Raritan, NJ) for 3 h under atmospheric conditions. SDE cold finger temperature was maintained at -4°C. Each extract was dried over 2 g of anhydrous sodium sulfate and concentrated to 0.5 mL under a gentle stream of purified nitrogen. Duplicate extractions were preformed for each sample.

Direct Solvent Extraction (DE). Whole saffron stigmas (2.5 g) were ground for 30 s in a spice mill (Model K7450C, Regal Ware, Inc., Kewaskum, WI) and then sieved through a no. 60 nylon mesh screen. Saffron powder (0.5 g), 10 µL of internal standard solution (as above), and 3 mL of redistilled dichloromethane were vigorously shaken for 5 min in a 16 x 125 mm screw capped test tube sealed with a PTFE-lined cap. The suspension was allowed to stand in the dark for 2 h, shaken again, and then centrifuged at 3,000xg for 30 min. After recovering the supernatant, the residue was extracted two more times with 3 mL of dichloromethane as described above. Extract was dried over 2 g of anhydrous sodium sulfate and concentrated to 0.5 mL under a gentle stream of purified nitrogen. Each saffron sample was extracted in duplicate.

Gas Chromatography-Mass Spectrometry (GC-MS). GC-MS system consisted of an HP 5890 Series II GC/HP 5972 mass selective detector (MSD, Hewlett-Packard,

Co., Palo Alto, CA.) One μL of each extract was injected (splitless mode; 30 s valve delay; 200°C injector temperature) into a capillary column (DB-wax or DB-5ms, 60 m length x 0.25 mm i.d. x 0.25 μm film thickness (d_f); J & W Scientific, Folson, CA). Helium was used as carrier gas at a constant flow rate of 0.96 mL/min. Oven temperature was programmed from 40°C to 200°C at a rate of 3°C/min with initial and final hold times of 5 and 60 min, respectively. MSD conditions were as follows: capillary direct interface temperature, 280°C; ionization energy, 70 eV; mass range, 33-350 a.m.u.; EM voltage, 1956 (Atune + 200V); scan rate, 2.2 scans/s. Each SDE or DE extract was analyzed in duplicate.

Aroma Extract Dilution Analysis (AEDA). GC-olfactometry (GC-O) system consisted of a Varian 3300 GC (Varian Instrument Group, Walnut Creek, CA) equipped with a flame ionization detector (FID) and sniffing port. Column effluent was split 1:1 between FID and sniffing port by using deactivated capillary columns (1 m length x 0.25 mm i.d.). Serial dilutions (1:3) were prepared from SDE and DE extracts using dichloromethane as diluent. Each dilution (1 μL) was injected into an FSOT column (DB-wax or DB-5ms, 30 m length x 0.32 mm i.d. x 0.25 μm d_f, J & W Scientific). GC conditions were same as GC-MS except that oven temperature was programmed at a rate of 6°C/min and with initial and final hold times of 5 and 30 min, respectively. FID and sniffing port were held at 250°C.

GC/O was performed by two trained panelists familiar with saffron aroma. Each panelist evaluated a dilution series for only one SDE or DE extract for each saffron sample. Because of the similarity between GC/O results for SDE extracts from all three saffron samples, and likewise DE extracts, final AEDA results were summarized by calculating average log_3(FD factors) (n = 6) for each type of extract.

Compound Identification and Quantitation. Compound identifications were based on comparison of GC retention indices (RI)(20), mass spectra, and aroma properties of unknowns with those of authentic standard compounds analyzed under identical experimental conditions. Tentative identifications were based on matching mass spectra of unknowns with those in the Wiley 138K mass spectral database (John Wiley and Sons, Inc., 1990) and literature or on matching RI values and aroma properties of unknowns with those of authentic standard compounds.

Relative concentrations of positively identified compounds were determined using their MS response factors compared with the internal standard. Response factors were determined by analyzing standard compounds at three levels under identical GC/MS conditions. The relative concentration of each tentatively identified compound was estimated from its peak area relative to the internal standard.

Results & Discussion

Volatile components of saffron. Typical total ion chromatograms of SDE and DE volatile extracts of saffron are shown in Figures 1 and 2, respectively. Quantitative results are presented in Table I. A total of 46 core volatile components were identified. Of these, 30 compounds were found common to both SDE and DE extracts. Safranal (2,6,6-trimethyl-1,3-cyclohexadiene-1-carboxaldehyde, no. 31) was the most abundant volatile component, followed by isophorone (3,5,5-trimethyl-2-cyclohexen-1-one, no. 26), and 2,6,6-trimethyl-2-cyclohexene-1,4-dione (no. 36). The

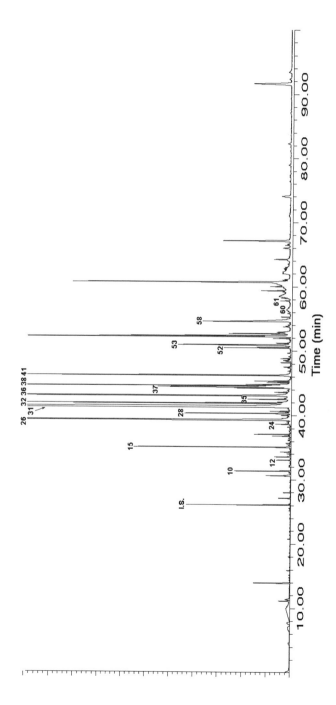

Figure 1. Total ion chromatogram of saffron volatiles isolated by simultaneous steam distillation-solvent extraction. Peak numbers correspond to those in Tables I and II and Figure 2.

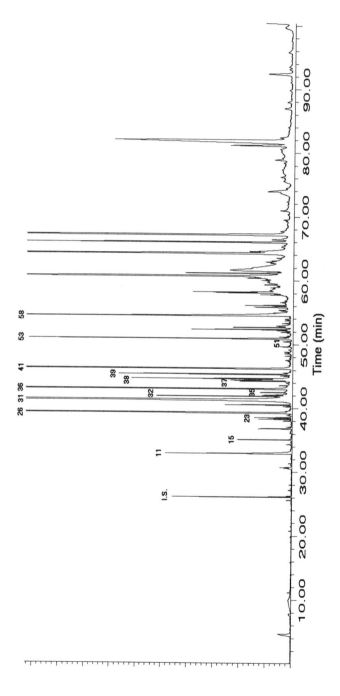

Figure 2. Total ion chromatogram of saffron volatiles isolated by direct solvent extraction. Peak numbers correspond to those in Tables I and II and Figure 1.

Table I. Volatile Compounds in Saffron

No.[a]	Compound	RI[b]		Conc. (μg/g)[c]	
		DB-wax	DB-5ms	SDE[d]	DE[e]
2	5-Tert-butyl-1,3-cyclopentadiene[f*]	966	844	6.0±1.9	2.0±0.2
3	2,3-Butanedione[g*]	975	<700	9.3±1.7	tr[h]
5	3-Hydroxy-2-butanone[g*]	1285	711	nd[j]	73±30
6	3-Heptanol (I.S.[k])	1295	899	- -	- -
10	3,5,5-Trimethyl-3-cyclohexen-1-one[f]	1412	1044	61±10	nd
11	Acetic acid[g]	1446	<700	nd	754±108
13	Megastigma-7,9,13-triene[f*]	1452	1262	11±2	nd
14	2-Furancarboxaldehyde[g*]	1463	832	23±9	nd
15	2-Methylene-6,6-dimethyl-3-cyclohexene-1-carboxaldehyde[i]	1498	1112	120±22	34±4
18	2,6,6-Trimethyl-3-oxo-3-cyclohexene-1-carboxaldehyde[i]	1541		11±1	26±3
19	Linalool[g*]	1546	1104	26±5	2.4±0.3
20	2-Methylpropanoic acid[f*]	1569		nd	9.4±1.6
21	5-Methyl-2-furancarboxaldehyde[g*]	1575		7.3±2.8	nd
22	2,3-Butanediol[g*]	1577	794	nd	21±2
23	Sulfinylbis methane[f*]	1582	844	nd	28±2

Continued on next page.

Table I Continued

No.[a]	Compound	RI[b]		Conc. (µg/g)[c]	
		DB-wax	DB-5ms	SDE[d]	DE[e]
24	Megastigma-4,6,8-triene isomer[f*]	1586	1323	13±5	nd
26	3,5,5-Trimethyl-2-cyclohexen-1-one (isophorone)[g]	1602	1129	1270±130	954±130
27	Megastigma-4,6,8-triene isomer[f*]	1620	1363	9.7±1.9	nd
28	Megastigma-4,6,8-triene isomer[f*]	1628	1365	80±18	7.4±1.1
29	Dihydro-2(3H)-furanone[f*]	1633		7.5±3.4	59±12
31	2,6,6-Trimethyl-1,3-cyclohexadiene-1-carboxaldehyde (safranal)[g]	1656	1210	5250±390	3480±300
32	2-Hydroxy-3,5,5-trimethyl-2-cyclohexen-1-one[f]	1668	1152	207±39	97±12
33	3-Methylbutanoic acid[g*]	1672	869	nd	90±13
35	4-Hydroxy-3,5,5-trimethyl-2-cyclohexen-1-one[i]	1683		55±21	23±6
36	2,6,6-Trimethyl-2-cyclohexene-1,4-dione[g]	1698	1151	1280±170	921±120
37	2-Hydroxy-4,4,6-trimethyl-2,5-cyclohexadien-1-one[i]	1734	1163	258±28	84±8
38	2,6,6-Trimethyl-2,4-cycloheptadien-1-one[f*]	1740	1229	212±18	128±15
39	2(5H)-Furanone[g]	1761	912	nd	400±55
41	2,2,6-Trimethyl-1,4-cyclohexanedione[f,i]	1785	1179	437±135	554±70
44	2-Phenethylacetate[g*]	1820	1258	4.5±0.7	4.4±0.4
45	Dihydro-ß-ionone[f*]	1839		7.4±1.0	3.6±0.6
46	Hexanoic acid[g*]	1849		nd	33±7
47	(E)-Geraniol[g*]		1849	11±4	nd
48	6,10-Dimethyl-5,9-undecadien-2-one isomer[f*]	1857		7.1±1.0	3.9±0.8

Continued on next page.

Table I Continued

No.[a]	Compound	RI[b]		Conc. (μg/g)[c]	
		DB-wax	DB-5ms	SDE[d]	DE[e]
49	Benzenemethanol[g*]	1880	1043	4.8±1.5	6.9±1.2
51	Sulfonylbis methane[f*]		1903	nd	7.5±0.9
52	2,4,6-Trimethylbenzaldehyde[f,i]	1904	1321	58±10	tr
53	2-Phenylethanol[g]	1917	1121	231±63	264±29
54	2,4,6-Tetramethyl-1-cyclohexene-1-carboxaldehyde[i]	1917	1313	N/A[l]	N/A
55	β-Ionone[g*]	1947	1484	8.6±1.1	4.7±1.1
56	2,6,6-Trimethyl-3-oxo-1,4-cyclohexadiene-1-carboxaldehyde[i]	1970	1312	70±12	42±7
58	2-Hydroxy-3,5,5-trimethyl-2-cyclohexen-1,4-dione[f,i]	2023	1240	281±130	378±53
60	2,6,6-Trimethyl-4-oxo-2-cyclohexen-1-carboxaldehyde[f*]	2083		37±24	tr
61	2,6,6-Trimethyl-5-oxo-1,3-cyclohexadiene-1-carboxaldehyde[i]	2127	1369	15±5	28±10
62	4-Hydroxy-2,6,6-trimethyl-2-cyclohexen-1-one[f*]	2141	1258	15±5	93±37
63	4-Hydroxy-2,6,6-trimethyl-3-oxo-1-cyclohexene-1-carboxaldehyde[i]	2152	1346	129±75	74±18
64	5-(Buta-1,3-dienyl)-4,4,6-trimethyl-1,5-cyclohexadien-1-ol[i]	2179	1501	40±13	39±10

[a]Peak numbers correspond to those in Figures 1 and 2 and Table II. [b]RI, retention index. [c]Average relative concentration ± standard deviation (n=12). [d]SDE, simultaneous steam distillation-solvent extraction. [e]DE, direct solvent extraction. [f]Compound tentatively identified by comparing its mass spectrum to Wiley 138K mass spectral database. [*]Compound not previously identified in saffron. [g]Compound positively identified as described in materials and methods. [h]tr, trace. [i]Compound tentatively identified by comparing its mass spectrum with published literature (11). [j]nd, not detected. [k]I.S., internal standard. [l]N/A, not available, peak could not be resolved.

concentrations of safranal presented in Table I were similar to published values (6). Likewise, the relative abundances (with respect to safranal) for prominent volatiles (e.g., nos. 15, 26, 31, 36, 38, 41, 53, and 58) were in reasonable agreement with previous studies (9,11).

Despite agreement with published literature, volatile profiles of SDE and DE extracts differed from each other with respect to levels of major volatiles as well as in types of compounds identified. DE allowed for the isolation of acidic components (nos. 11, 20, 34, and 46), while no acids were detected in SDE extracts. On the other hand, SDE extracts contained higher levels of major saffron volatiles (e.g., nos. 26, 31, and 36). Higher levels of these components may have been a result of thermally induced hydrolysis of their glucoside precursors during SDE (9). Furthermore, it also is likely that other thermally generated compounds were formed during SDE. For example, the sugar breakdown products, 2-furancarboxyaldehyde (no. 14) and 5-methyl-2-furancarboxaldehyde (no. 21), were only detected in SDE extracts. SDE extracts also contained four megastigmatriene isomers (nos. 13, 24, 27, and 28), which have been previously reported as volatile constituents of starfruit (21).

In addition to the above mentioned compounds, many others were identified for the first time as saffron components (see compounds indicated by an asterisk in Table I). Of particular interest and importance were the following compounds detected by both GC-MS and GC-O: 2,3-butanedione (no. 3), acetic acid (no. 11), linalool (no. 19), and 3-methylbutanoic acid (no. 33). Levels of these compound were comparatively low with respect to previously identified components.

Aroma-active components of saffron. The characteristic aroma of saffron has been previously defined in the literature as "sweet, spicy, floral odor with a fatty, herbaceous undertone (1)". Despite quantitative and qualitative differences between SDE and DE extracts, both had distinct saffron-like aromas, although their aromas were not identical. DE extracts had sweet, spicy, and floral notes and were representative of dried saffron; whereas, SDE extracts had nutty, cooked rice- and hay-like notes but still maintained a definite saffron-like quality.

A total of 25 aroma-active compounds were consistently detected by GC-O and AEDA in SDE and DE extracts (Table II). More compounds (23) were detected in SDE extracts than DE extracts (22), with 18 common to both extracts. Aromas detected during GC-O could be grouped into the following general categories: saffron-like (nos. 10, 30, 31, 34, 37, and 57), fatty, stale, bitter (nos. 16, 17, 42, 43, and 50), sweet, floral (nos. 19, 25, 53, and 59), sour (nos. 1, 11, and 33), nutty, cooked (nos. 8, 9, and 12), earthy, plastic (nos. 4 and 7), buttery (no. 3) and green onion (no. 40). In general, compounds that were not common to both extracts had low $\log_3$(FD-factors) (<1). For compounds detected in both extracts, most had higher $\log_3$(FD-factors) in SDE extracts. This was especially true for nutty, cooked aromas which may have been thermally generated during SDE extraction.

Compounds having high $\log_3$(FD-factors) (>2) in both extracts were described as saffron, dried hay (unknown, nos. 34 and 37), saffron, tea (safranal, no. 31) and cooked rice, baked bread (no. 9). Compound no. 37 had the highest $\log_3$(FD-factor) in both extracts. Interestingly, this compound and no. 34 were detected at only high dilutions (>1:9), i.e., these compounds were not detected during GC/O of concentrated extracts. This phenomenon was observed previously (14) and may have been due to

Table II. Results of Aroma Extract Dilution Analysis

No.[a]	Compound Name	RI[b] DB-wax	RI[b] DB5-ms	Aroma Description[c]	Av. log$_3$(FD factor)[d] SDE[e]	Av. log$_3$(FD factor)[d] DE[f]
1	unknown	933	605	sour, dark chocolate	<1	<1
3	2,3-Butanedione[g]*	974	614	buttery, cream cheese	<1	<1
4	unknown	1096	775	plastic water bottle	<1	nd[h]
7	1-Octen-3-one[i]*	1299	978	mushroom, earthy	2.0	1.17
8	2-Acetyl-1-pyrroline[i]*	1336	921	nutty, popcorn	2.83	<1
9	unknown	1374	866	cooked rice, baking bread	3.5	2.0
10	3,5,5-Trimethyl-3-cyclohexen-1-one[j]	1405	1042	saffron, floral, hay	1.33	<1
11	Acetic acid[g]*	1445	605	vinegar, acidic	nd	<1
12	3-(Methylthio)propanal[i]*	1452	906	baked potato	1.67	nd
16	unknown	1506		stale, bitter, dried hay	1.33	<1
17	unknown	1536		stale, bitter	1.33	nd
19	Linalool[g]*	1549	1096	floral, honeysuckle	1.67	<1
25	(E,Z)-2,6-Nonadienal[i]*	1589	1152	sweet, cucumber	1.0	<1

Continued on next page.

Table II Continued

No.[a]	Compound Name	RI[b]		Aroma Description[c]	Av log₃(FD factor)[d]	
		DB-wax	DB5-ms		SDE[e]	DE[f]
30	unknown	1638		saffron, dried hay	2.0	1.17
31	2,6,6-Trimethyl-1,3-cyclohexadiene-1-carboxaldehyde (safranal)[g]	1645	1203	saffron, tea	5.33	4.0
33	3-Methylbutanoic acid[g]*	1672	840	rotten, sour, dried fruit	2.0	1.67
34	unknown	1682		saffron, dried hay	2.17	4.5
37	2-Hydroxy-4,4,6-trimethyl-2,5-cyclohexadien-1-one[k]	1734	1159	saffron, stale, dried hay	5.5	5.5
40	unknown	1775		green onion	1.83	1.33
42	unknown	1804		stale, soapy	nd	<1
43	(E,E)-2,4-Decadienal[i]*	1817	1312	fatty, fried fat	1.0	1.0
50	unknown	1886		fruity, stale	<1	nd
53	2-Phenylethanol[g]	1920	1115	floral, rose	<1	1.33
57	unknown	1980		floral, rose, saffron	3.17	<1
59	4-Hydroxy-2,5-dimethyl-3(2H)-furanone[i]*	2049	1060	cotton candy, strawberries	nd	<1

[a]Numbers correspond to those in Figures 1 and 2 and Table I. [b]Retention indices calculated from GC-O results. [c]Aroma quality as perceived during GC-O. [d]Average log₃(FD factor) of 6 replicates. [e]SDE, simultaneous steam distillation-solvent extraction. [f]DE, direct solvent extraction. [g]Compound positively identified as described under materials and methods. *Compound not previously identified in saffron. [h]nd, not detected by GC-O. [i]Compound tentatively identified by comparing its RI and aroma quality with standard compound. [j]Compound tentatively identified by comparing its mass spectrum to Wiley 138K mass spectral database. [k]Compound tentatively identified by comparing its mass spectrum with published literature (11).

cross adaptation of the panelists to an earlier eluting compound (e.g., no. 31 or 34) or masking by an earlier eluting aroma. The peak and mass spectrum corresponding to compound no. 37 could be clearly identified since the compound eluted before safranal (no. 31) on DB-5ms columns and eluted after safranal on DB-wax columns. Based on its mass spectrum (Figure 3), compound no. 37 was tentatively assigned the molecular formula of $C_9H_{12}O_2$ and structure of 2-hydroxy-4,4,6-trimethyl-2,5-cyclohexadien-1-one. The proposed structure is supported by mass spectral data of Rödel and Petrzika (*11*). Compounds nos. 9 and 34 could not be identified due to inadequate MS signals at their corresponding RI values. The $\log_3$(FD-factors) for safranal (no. 31) were nearly as high as those for no. 37, indicating its importance to saffron aroma. This finding was not surprising since safranal has long been regarded as the character-impact component of saffron (*10-11*).

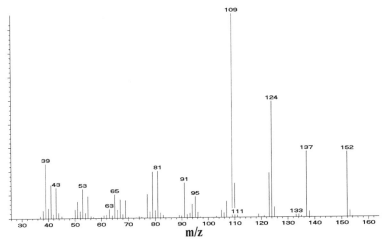

Figure 3. Mass spectrum of peak no. 37 in Figures 1 and 2 and Tables I and II.

There were a number of compounds with moderate aroma intensities [$\log_3$(FD factors) > 1 in either SDE or DE or both]. Positively identified compounds within this group were described as floral, honeysuckle (linalool, no. 19), rotten, sour, dried fruit (3-methylbutanoic acid, no. 33) and floral, rose (2-phenylethanol, no. 53). Linalool and 2-phenylethanol are common glycosidically bound constituents of plants (*22-23*). Due to low MS signals several compounds were identified based on their RI values and aroma properties (i.e., compared against standard compounds). Their aromas were described as mushroom, earthy (1-octen-3-one, no. 7), nutty, popcorn (2-acetyl-1-pyrroline, no. 8), baked potato [3-(methylthio)propanal, no. 13], sweet, cucumber [(E,Z)-2,6-nonadienal, no. 25], and fatty, fried fat [(E,E)-2,4-decadienal], no. 43). These compounds impart characteristic aromas to many foods and are known to have low thresholds. For example, the lipid oxidation products nos. 7, 25 and 43 (*24*) have threshold values of 0.1 (*25*), 0.1 (*26*), and 0.07 µg/L (*27*), respectively. Similarly, thermally generated compounds such as no. 8, a Maillard reaction product (*28*), and no. 13, a Strecker aldehyde from methionine (*29*), have thresholds of 0.1

(*27*) and 0.2 µg/L (*30*), respectively. A compound with a saffron, floral, hay-like aroma was tentatively identified by its mass spectrum as 3,5,5-trimethyl-3-cyclohexen-1-one (no. 10). Unidentified compounds within this group were described as stale, bitter (nos. 16 and 17), green onion (no. 40), and floral, rose, saffron (no. 57).

Compounds considered to have only minor roles in saffron aroma [$\log_3$(FD factors) < 1] were described as sour, dark chocolate (unknown, no. 1), buttery, cream cheese (2,3-butanedione, no. 3), plastic water bottle (unknown, no. 4), vinegar, acidic (acetic acid, no. 11), stale, soapy (unknown, no. 42), fruity, stale (unknown, no. 50) and cotton candy, strawberries [4-hydroxy-2,5-dimethyl-3(2*H*)-furanone, no. 59]. With its low threshold compound no. 59 is an important component of several foods (*18,31*).

Conclusion

Through use of AEDA and two extraction techniques it was possible to indicate important aroma-active components in saffron. Aromas contributed by 2-hydroxy-4,4,6-trimethyl-2,5-cyclohexadien-1-one (tentatively identified), safranal, and an unidentified compound (saffron, dried hay aroma) were predominant in saffron. However, it also was apparent that other components contribute to saffron aroma, some of which are thermally generated during cooking of saffron. Many compounds were identified for the first time as constituents of saffron. Results of this study may be useful in development of analytical strategies for monitoring saffron flavor quality.

Acknowledgements

Approved for publication by the Mississippi Agricultural and Forestry Experiment Station as manuscript No. PS-8919. Support for this study was provided by the Mississippi Agricultural and Forestry Experiment Station.

Literature Cited

1. *Fenaroli's Handbook of Flavor Ingredients,* Burdock, G.A. Ed.; CRC Press: Boca Raton, FL, 1995; Vol 1, 3rd ed., pp 247-248.
2. Sampathu, S.R.; Shivashankar, S.; Lewis, Y.S. *Crit. Rev. Food Sci. Nutr.* **1984,** *20,* 123-157.
3. Oberdieck, R. *Dtsch. Lebensm.-Rundsch.* **1991,** *87,* 246-252.
4. Castellar, M.R.; Montijano, H.; Manjón, A.; Iborra, J.L. *J. Chromatogr., A* **1993,** *648,* 187-190.
5. Iborra, J.L.; Castellar, M.R.; Cánovas, M.; Manjón, A. *J. Food Sci.* **1992,** *57,* 714-731.
6. Sujata, V.; Ravishankar, G.A.; Venkataraman, L.V. *J. Chromatogr.* **1992,** *624,* 497-502.
7. Tarantilis, P.A.; Polissiou, M.; Manfait, M. *J. Chromatogr., A* **1994,** *664,* 55-61.
8. Tarantilis, P.A.; Tsoupras, G.; Polissiou, M. *J. Chromatogr., A* **1995,** *669,* 107-118.
9. Zarghami, N.S.; Heinz, D.E. *Phytochem.* **1971,** *10,* 2755-2761.
10. Narasimham, S.; Chand, N.; Rajalakshmi, D. *J. Food Qual.* **1992,** *15,* 303-314.

11. Rödel, W.; Petrzika, M. *J. High Res. Chromatogr.* **1991**, *14*, 771-774.
12. Acree, T. In *Flavor Measurement*; Ho, C.-T., Manley, C.H., Eds.; Dekker: New York, 1993; Vol. 1, Chapter 4.
13. Grosch, W. *Trends in Food Sci Technol.* **1993**, *4*, 68-73.
14. Abbott, N.; Etievant, P.; Issanchou, S.; Langlois, D. *J. Agric. Food Chem.* **1993**, *41*, 1698-1703.
15. Guichard, H.; Guichard, E.; Langlois, D.; Issanchou, S.; Abbott, N. *Z. Lebensm. Unters. Forsch.* **1995**, *201*, 344-350.
16. Cadwallader, K.R.; Tan, Q.; Chen, F.; Meyers, S.P. *J. Agric. Food Chem.* **1995**, *43*, 2432-2437.
17. Chung, H.Y.; Cadwallader, K.R. *J. Agric. Food Chem.* **1994**, *42*, 2867-2869.
18. Guth, H.; Grosch, W. *Lebensm.-Wiss. u.-Technol.* **1993**, *26*, 171-177.
19. Schieberle, P.; Grosch, W. *Z. Lebensm. Unters. Forsch.* **1987**, *185*, 111-113.
20. van den Dool, H.; Kratz, P.D. *J. Chromatogr.* **1963**, *11*, 463-471.
21. MacLeod, G.; Ames, J.M. *Phytochem.* **1990**, *29*, 165-172.
22. Vasserot, Y; Arnaud, A.; Galzy, P. *Acta Biotechnol.* **1995**, *15*, 77-95.
23. Shu, C.-K.; Lawrence, B.M. *J. Agric. Food Chem.* **1994**, *42*, 1732-1733.
24. *Lipid Oxidation in Food*, St. Angelo, A.J.,; ACS Symposium Series No. 500.; American Chemical Society: Washington D.C., 1992.
25. Whitfield, F.B.; Freeman, D.J.; Last, J.H. *Austr. J. Chem.* **1982**, *35*, 373-383.
26. Forss, D.A.; Dunstone, E.A.; Ramshaw, E.H.; Stark, W. *J. Food Sci.* **1962**, *27*, 90-93.
27. Buttery, R.G.; Seifer, R.M.; Ling, L.C. *J. Agric. Food Chem.* **1988**, *36*, 1006-1009.
28. Schieberle, P. *Z. Lebensm. Unters. Forsch.* **1990**, *191*, 206-209.
29. Forss, D.A. *J. Dairy Res.* **1979**, *46*, 691-706.
30. Guadagni, D.; Buttery, R.G.; Turnbaugh, J.G. *J. Sci. Food Agric.* **1972**, *23*, 1435-1444.
31. Meyerl, F.; Regula, N.; Thomas, A. *Phytochem.* **1989**, *28*, 631-633.

Chapter 8

The Characterization of Volatile and Semivolatile Components in Powdered Turmeric by Direct Thermal Extraction Gas Chromatography–Mass Spectrometry

Richard D. Hiserodt[1,2], Chi-Tang Ho[1], and Robert T. Rosen[2]

[1]Department of Food Science and [2]Center for Advanced Food Technology, Cook College, Rutgers, The State University of New Jersey, New Brunswick, NJ 08903–0231

Five commercial powdered turmeric samples were analyzed to identify volatile and semi-volatile components. This analysis did not include the non-volatile curcuminoids. Structural information was obtained by direct thermal extraction gas chromatography-mass spectrometry (GC-MS) using the electron ionization (EI) mode. Semi-quantitative values are also reported.

Turmeric belongs to the family *Zingiberaceae* along with the other noteworthy members ginger, cardamom, and galangal root. It belongs to the genus *Curcuma* which consists of hundreds of species of plants that grow from rhizomes, underground rootlike stems. Economically, the most important species is *domestica*. Turmeric is grown in warm, rainy regions of the world such as China, Indonesia, India, Jamaica, and Peru (*1*).

Turmeric has been used since early times to cure everything from leprosy to the common cold but is probably best known for its properties as a carminative. It is used to provide aroma and taste to foods and as a dye for fabric. Its properties as a dye are poor because turmeric is not light stable. Turmeric is also used in some cultures as a cosmetic, to lighten the skin.

The primary uses of turmeric today are as a component in curry, as a coloring in a variety of dried and frozen foods and as a powdered spice. The powdered spice is prepared by harvesting the fresh rhizomes and boiling them to gelatinize the starch and disperse the color. The rhizomes are dried in the sun for 10 to 15 days and then ground into a powder (*1*). The work presented in this report is on the analysis of volatile and semi-volatile components in powdered turmeric. This work does not include the non-volatile curcuminoids

The primary focus of turmeric research today is based on its properties as an antioxidant (*2,3*) and as an anticarcinogen (*4,5,6*). The antioxidant properties of turmeric are based on the ability of curcumin to form complexes with metals and to form a resonance stabilized free radical. Curcumin, along with the other curcuminoids, demethoxycurcumin and *bis*demethoxycurcumin, are nonvolatile components of turmeric that impart its distinctive orange-yellow color. Curcumin complexes with metals, such as copper and nickel, that initiate or catalyze free radical oxidation. This property enables curcumin to act synergistically with other antioxidants in inhibiting free radical oxidation (7).

The bond dissociation energy for the phenolic hydroxyl groups in curcumin is low and a phenoxide free radical is easily formed (Figure 1). This free radical is resonance stabilized. The odd electron can be delocalized through the entire molecule because this molecule is a completely conjugated system. Figure 2 shows curcumin acting as a free radical scavenger, reacting with a peroxy free radical, forming a hydroperoxide and the curcumin free radical. The other property of curcumin that makes it a good antioxidant is that once the free radical is formed it will not go on to react with unsaturated fatty acids and initiate or propagate oxidation reactions.

Turmeric has been shown to inhibit the initiation and progression of cancer. Azuine and Bhide (4) demonstrated that turmeric inhibited forestomach tumors induced by benzo[α]pyrene (BP) and skin tumors induced by 7,12- dimethyl-benz[α]anthracene (DMBA) in mice. These studies were conducted at the 2% and 5% levels. Benzo[α]pyrene, which is a component in cigarette smoke and barbecue foods, is a procarcinogen that is oxidized during Phase I oxidation in the liver to an ultimate carcinogen (Figure 3). The ultimate carcinogen formed is a good alkylating agent and can react with DNA to form a mutation. Turmeric has been shown to inhibit tumors formed by this mechanism by inhibiting the P450 monooxygenase system. Turmeric also increases levels of glutathione and glutathione S-transferase activity. Glutathione is an endogenous nucleophillic chemical that can react with ultimate carcinogens formed during Phase I oxidation without forming tumors. Turmeric has also been shown to inhibit the progression of cancer by inhibiting ornithine decarboxylase. This enzyme decarboxylates ornithine, an endogenous amino acid, forming a polyamine. This polyamine stimulates cell and consequently tumor growth.

In another study, Mukundan, *et al* (5) showed that as little as 0.1% turmeric in the diet significantly removed BP-DNA adducts or inhibited the binding of DNA with BP formed in rat liver. They also observed 0.03% curcumin to be more effective in inhibiting BP-DNA adducts. They did not propose a mechanism for this observation.

Background

Preliminary work by these authors involved the analysis of components in the methanol extract of powdered turmeric using thermospray and particle beam liquid chromatography-mass spectrometry (LC-MS). Thermospray is a soft ionization technique producing protonated and deprotonated molecular ions, providing molecular weight information about analytes. The thermospray LC-MS chromatogram for the methanol extract of powdered turmeric is shown in Figure 4. The curcuminoids were identified based on their molecular weight. Also detected were three major late eluting components as well as numerous minor components. The molecular weights for these components were obtained from the thermospray mass spectra. Particle beam LC-MS was attempted to obtain structural information from EI-mass spectra, for the

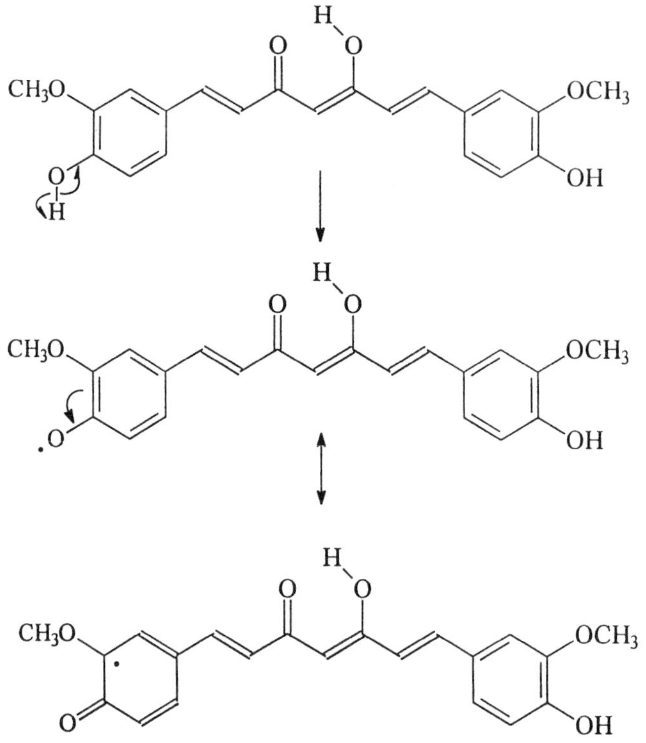

Figure 1. Resonance Stabilization of the Curcumin Free Radical

$$L-OO\bullet \ + \ HO-C \longrightarrow L-OOH \ + \bullet O-C$$

Peroxy   Curcumin   Curcumin
Free Radical   Hydroperoxide   Free Radical

$$L-H \ + \ \bullet O-C \ \longrightarrow\!\!\!\!\!X \longrightarrow$$

Unsaturated   Curcumin
Fatty Acid   Free Radical

Figure 2. Reaction Mechanism for Curcumin Acting as a Free Radical
Scavenger

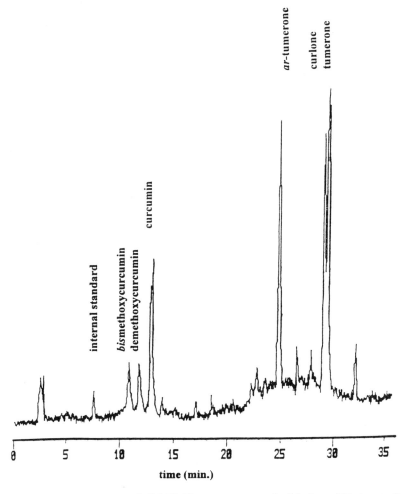

Figure 3. Phase I Oxidation of Benzo[α]pyrene

Figure 4. Thermospray LC-MS Chromatogram of a Methanol Extract of Powdered Turmeric

components detected by thermospray. The problem encountered with particle beam LC-MS was that none of the late eluting peaks detected by thermospray were observed. This phenomenon was attributed to the volatility of the components that were lost during removal of the mobile phase in the particle beam interface. Structural information for the volatile components in the methanol extract of turmeric was then obtained by direct thermal extraction GC-MS.

Materials

Samples of powdered turmeric, consisting of 5 different commercial labels, were purchased locally. Chromosorb WHP (80/100 m) was purchased from Supelco, Inc., Bellefonte, PA, USA. Chromosorb WHP was used to dilute the powdered turmeric and was conditioned at 300 °C for 2 hrs. prior to use. d_8-Naphthalene (98+ atom %D) was purchased from Aldrich Chemical Company, St. Louis, MO, USA for use as an internal standard. The internal standard solution was prepared by diluting 10.0 mg of d_8-naphthalene with 10.0 mL methanol. A 60 m x 0.32 mm (i.d.) df = 0.25 μm, DB-1 capillary GC column was purchased from J&W Scientific, Folsom, CA, USA.

Experimental

There are many alternatives to direct thermal extraction methodology. For instance compounds could be isolated using the Likens Nikerson system, which provides for steam distillation of the aqueous sample with ether extraction on a continuous basis in one apparatus. The disadvantages of this system are that the steam distillation utilizes high temperatures (100°C) which may decompose labile compounds or cause other thermal induced reactions to occur in the product being analyzed thereby complicating the data interpretation. Additionally, the ether extract from the Likens Nikerson system needs to be concentrated prior to GC-MS analysis and there is always a risk of losing highly volatile components from the sample during the concentration step which is usually performed in a spinning band still or Kuderna Danish concentrator.

   Another alternative is to use vacuum steam distillation combined with cryotrapping. In this technique the sample is steam distilled under reduced pressure so that lower temperature isolation is permissible. The steam distillate is then condensed in a series of four or more cryotraps cooled with liquid nitrogen and or dry ice acetone slurries. The major disadvantage of this technique is that the first several cryotraps rapidly become filled with ice from the large volume of distillate. After thawing the traps, the water and traps must be extracted with organic solvent such as ether or methylene chloride to isolate the volatile organics. There are additional problems with extraction efficiency for polar organics that are not easily extracted from the water. This results in large volumes of organic solvent that must be concentrated, once again risking the loss of highly volatile compounds in the process. The extracting solvents used must be of ultra high purity. Frequently the extracting solvents must be distilled before use because concentrating large volumes of solvent will concentrate solvent impurities as well. Blank runs of solvent, with no sample, are often needed to identify artifacts and impurities arising from the solvent concentration step. Extraction experiments are time consuming and costly since elaborate glassware setups, vacuum systems, and concentration equipment are employed.

Direct thermal extraction involves the extraction of volatile and semi-volatile components from solid samples using heat (*8*). Since volatile components are desorbed directly onto the head of a GC capillary column at subambient conditions, samples should contain less than 5-10% water to prevent ice formation at the head of the column. Direct thermal extraction is a single step process that involves minimal or no sample preparation making it advantageous over other extraction techniques. Additional benefits are no solvent disposal and no artifact peaks in the sample chromatogram from the concentration of large volumes of extracting solvent.

The instrument used for this experiment was the Short Path Thermal Desorption (SPTD) system TD-3. This is a commercial system developed jointly by the Center for Advanced Food Technology (CAFT) at Rutgers University and Scientific Instrument Services, Inc. (SIS), Ringoes, NJ, USA. The TD-3 was connected to a Varian 3400 GC (Sugarland, TX, USA).

Powdered turmeric samples were prepared for quantitative analysis by thoroughly mixing 20.0 mg with 180.0 mg of Chromosorb WHP (80/100 m). Fifteen (15.0) mg of this mixture was weighed into a 10.2 cm x 4 mm (i.d.) glass lined stainless steel thermal desorption tube (Scientific Instrument Services, Inc., Ringoes, NJ, USA). Silanized glass wool was added to both ends to contain the sample. Both the glass wool and thermal desorption tube were conditioned at 300 °C for 2 hrs. prior to use. Five (5.0) µL of d_8-naphthalene solution (1.0 mg/ mL CH_3OH) was spiked into each sample. This was equivalent to 3,333 ppm based on the weight of turmeric in the turmeric-Chromosorb mixture. A stream of nitrogen (80 mL/min.) at room temperature was used to purge the desorption tube (30 min.) of methanol from the internal standard solution.

The thermal desorption tube has a needle at one end and is threaded at the other for connection to the SPTD system. When an injection is made, the thermal desorption tube needle pierces the GC injection port septum. Heating blocks close around the thermal desorption tube and provide rapid heating. Helium passes through the desorption tube at 1 mL/min. and the volatile components are desorbed onto a DB-1 capillary column held at -20 °C with dry ice (see Figure 5).

In this analysis the volatile components were trapped onto the head of the capillary column at subambient temperature (-20 °C) and separated by temperature programming the column from -20 to 150 °C at 20 °C/ min. followed by 150 to 280 °C at 5 °C/ min. Quantitative results were obtained using a flame ionization detector (FID). Structural information was obtained using a Finnigan MAT 8230 magnetic sector mass spectrometer, operating in the EI-mode, and interfaced to a SS300 data system (San Jose, CA, USA). The details of the instrument conditions are listed in Table I.

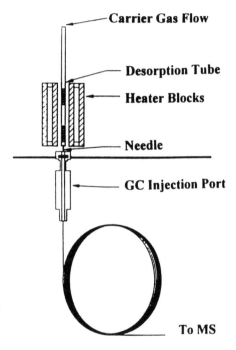

Figure 5. Schematic of the Short Path Thermal Desorption System

Table I. Instrument Conditions for the Analysis of Powdered Turmeric

Instrument: Short Path Thermal Desorption System TD-3

Carrier (He)	1 mL/min.
Heating block temp.	220 °C
Desorption time	5 min.

Instrument: Varian 3400 GC

Column	DB-1 60 m x 0.32 mm (i.d.) capillary column, df = 0.25 μm
Temp. prog.	-20 to 150 °C @ 20 °C/ min., 150 to 280 °C @ 5 °C/ min.
Inj. temp.	220 °C
FID temp.	325 °C

Instrument: Finnigan MAT 8230 Mass Spectrometer

Masses scanned	35-350 amu
EI-mode	70 eV @ 1mA
GC-MS interface line	280 °C
MS inlet temp.	240 °C
Ion source temp.	280 °C

Results and Discussion

Figure 6 shows a GC chromatogram of turmeric powder obtained by direct thermal extraction. The major late eluting peaks in this chromatogram were correlated with the late eluting peaks detected by thermospray LC-MS based on molecular weight and relative intensity.

The data obtained by GC using a flame ionization detector were used to obtain semi-quantitative results for the volatile components in five turmeric powders. Calculations were based on comparing the response of each component to the response of d_8-naphthalene used as an internal standard. The following equation was used for this calculation. Results are listed in Table II.

$$ppm = [(ISTDWT_{\mu g} \times AREA_{SPL}) / (AREA_{STD} \times SPLWT_g)] \times PERCENT]$$

where

ISTDWT is the weight of the internal standard added to each sample in μg

$AREA_{SPL}$ is the peak area for the component of interest

$AREA_{STD}$ is the peak area for the internal standard

SPLWT is the weight of the turmeric-Chromosorb mixture weighed into the desorption tube in grams

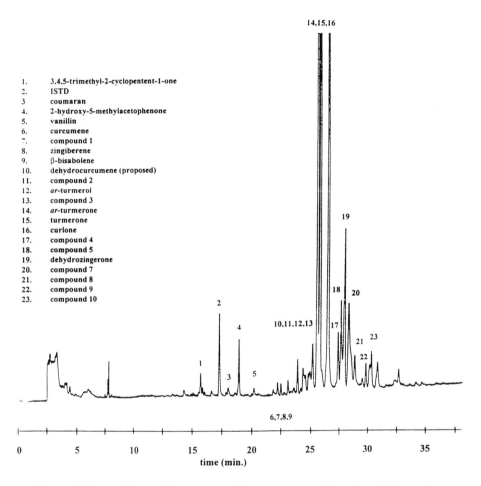

Figure 6. Direct Thermal Extraction Chromatogram of Powdered Turmeric

Table II. Semi-quantitative Values for Volatile and Semi-volatile Components in Powdered Turmeric

Retention time (min.)	Kovat's Index[a]	Assignment	MW	Powder 1 (ppm)	Powder 2 (ppm)	Powder 3 (ppm)	Powder 4 (ppm)	Powder 5 (ppm)
15.71	747.85	3,4,5-trimethyl-2-cyclopentene-1-one	140	1000	674	519	478	866
17.28	779.74	ISTD (internal standard)	136	4000	4660	4760	3290	2580
18.04	794.44	coumaran	120	532	327	129	98	231
18.96	809.00	2-hydroxy-5-methylacetophenone	150	2590	1150	1183	848	1650
20.23	826.54	vanillin[b]	152	411	148	229	272	245
22.25	852.29	curcumene	204	544	601	895	370	347
22.55	855.92	compound 1[c]	204	537	295	292	315	296
22.83	859.26	zingiberene	204	107	137	173	98	108
23.17	863.26	β-bisabolene	204	632	495	679	435	549
24.00	872.78	dehydrocurcumene (proposed)	200	1710	647	690	707	1180
24.46	877.92	compound 2	216	1440	611	754	924	1240
25.05	884.37	ar-turmerol	218	1510	327	822	533	1340
25.28	886.84	compound 3	204	3490	1340	1230	1590	2430
25.68	891.09	ar-turmerone	216	59500	25700	25450	29200	33200
25.89	893.29	turmerone	218	45700	13900	6420	20500	25600
26.59	900.64	curlone	218	48000	15900	13100	23800	26900
27.47	911.74	compound 4	216	3470	906	1250	1480	2300
27.72	914.83	compound 5	216	5950	1620	963	1200	ND
28.05	918.86	compound 6	218	11400	3620	3510	4250	7960
28.40	923.09	dehydrozingerone	192	10700	2640	3890	3930	5830
28.94	929.51	compound 7	234	3320	632	467	1370	2460
29.95	941.20	compound 8	232	1470	432	301	978	939
30.447	946.73	compound 9	216	2370	780	1740	946	2300
30.98	952.72	compound 10	234	2310	464	878	1480	2060

[a]See Majlat et al (16)
[b]Possible decomposition product (17)
[c]See Figure 10

Source: Reprinted from ref. 18. Copyright 1996 with kind permission from Elsevier Science - NL, Sara Burgerhartstraat 25, 1055 KV Amsterdam, The Netherlands.

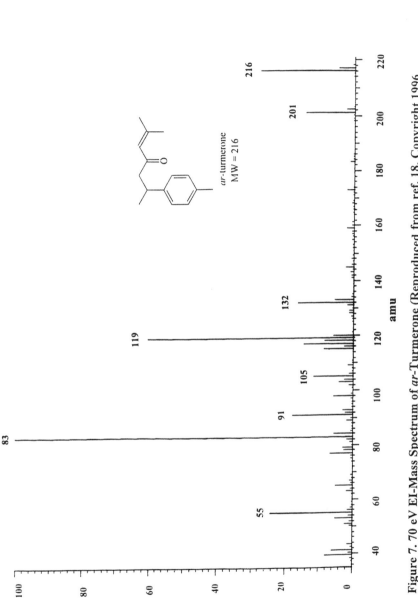

Figure 7. 70 eV EI-Mass Spectrum of *ar*-Turmerone (Reproduced from ref. 18. Copyright 1996 with kind permission from Elsevier Science - NL, Sara Burgerhartstraat 25, 1055 KV Amsterdam, The Netherlands.)

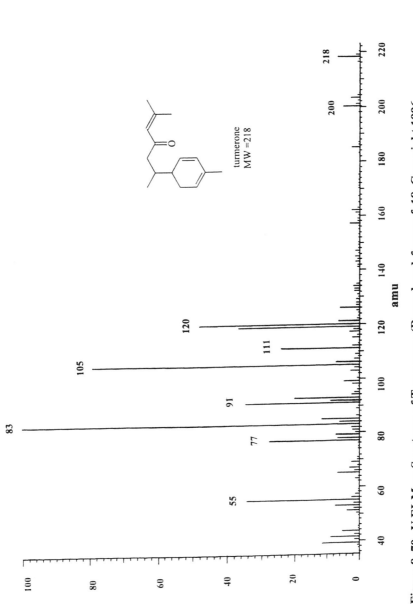

Figure 8. 70 eV EI-Mass Spectrum of Turmerone (Reproduced from ref. 18. Copyright 1996 with kind permission from Elsevier Science - NL, Sara Burgerhartstraat 25, 1055 KV Amsterdam, The Netherlands.)

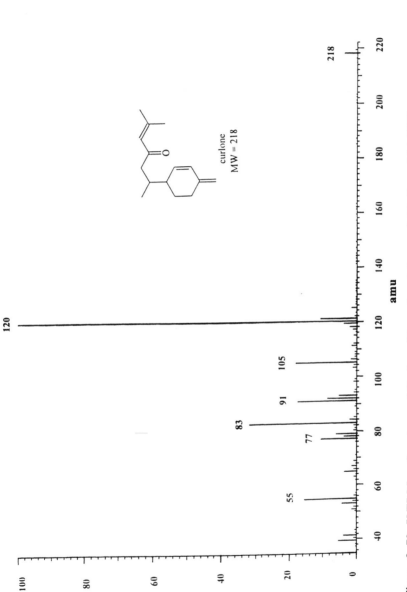

Figure 9. 70 eV EI-Mass Spectrum of Curlone (Reproduced from ref. 18. Copyright 1996 with kind permission from Elsevier Science - NL, Sara Burgerhartstraat 25, 1055 KV Amsterdam, The Netherlands.)

Table III. Eight Most Abundant Ions and Intensities for the Minor Volatile and Semi-volatile Components in Powdered Turmeric

Component	m/z (Intensity)							
Compound 1 MW=204 (2%)	41 (100%)	119 (75%)	93 (56%)	91 (38%)	69 (36%)	77 (34%)	39 (32%)	43 (26%)
Dehydrocurcumene MW=200 (4%)	119 (100%)	91 (22%)	120 (14%)	105 (12%)	41 (11%)	27 (10%)	43 (8%)	83 (7%)
Compound 2 MW=216 (4%)	119 (100%)	71 (38%)	43 (31%)	41 (29%)	91 (20%)	39 (18%)	132 (16%)	117 (12%)
ar-turmerol MW=218 (4%)	119 (100%)	85 (32%)	120 (16%)	117 (14%)	91 (12%)	157 (10%)	200 (8%)	185 (6%)
Compound 3 MW=204 (8%)	120 (100%)	41 (52%)	69 (50%)	91 (48%)	55 (47%)	93 (46%)	119 (44%)	105 (30%)
Compound 4 MW=220 (6%)	41 (100%)	137 (62%)	95 (46%)	110 (43%)	135 (41%)	55 (38%)	109 (34%)	67 (32%)
Compound 5 MW=216 (1%)	119 (100%)	83 (76%)	55 (27%)	91 (21%)	43 (16%)	39 (14%)	41 (13%)	120 (12%)
Compound 6 MW=218 (26%)	83 (100%)	135 (67%)	55 (53%)	123 (50%)	41 (40%)	107 (34%)	67 (32%)	91 (28%)
Compound 7 MW=234 (3%)	83 (100%)	55 (50%)	93 (40%)	114 (38%)	121 (33%)	91 (32%)	77 (30%)	41 (26%)
Compound 8 MW=232 (14%)	83 (100%)	199 (52%)	135 (46%)	55 (32%)	91 (21%)	39 (20%)	41 (15%)	43 (15%)
Compound 9 MW=216 (3%)	118 (100%)	83 (64%)	55 (30%)	117 (21%)	91 (18%)	136 (15%)	119 (14%)	39 (13%)
Compound 10 MW=234 (3%)	83 (100%)	55 (65%)	137 (55%)	110 (50%)	39 (38%)	41 (35%)	43 (28%)	95 (25%)

Source: Reprinted from ref. 18. Copyright 1996
with kind permission from Elsevier Science - NL, Sara Burgerhartstraat 25, 1055 KV Amsterdam, The Netherlands.

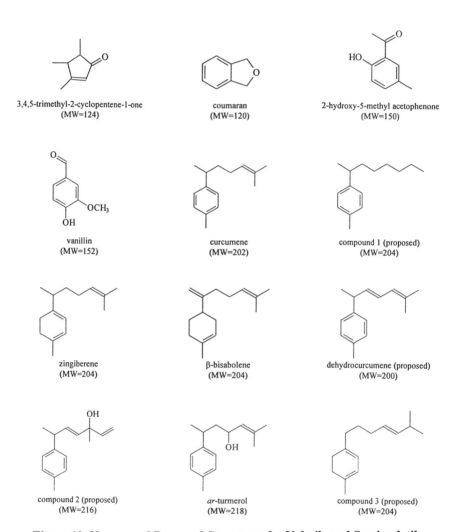

Figure 10. Known and Proposed Structures for Volatile and Semi-volatile Components in Turmeric

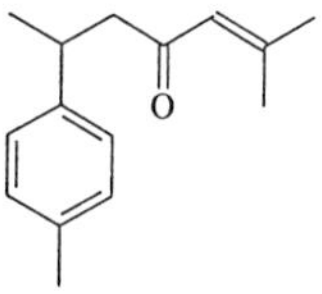

ar-turmerone
(MW=216)

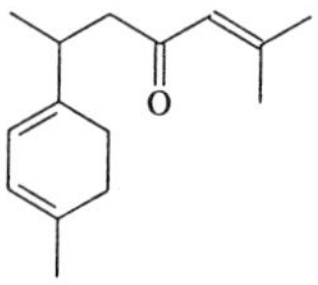

turmerone
(MW=218)

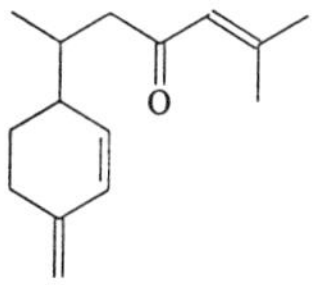

curlone
(MW=218)

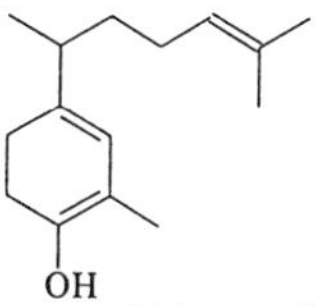

compound 4 (proposed)
(MW=220)

Unidentified
Component

compound 5
(MW=216)

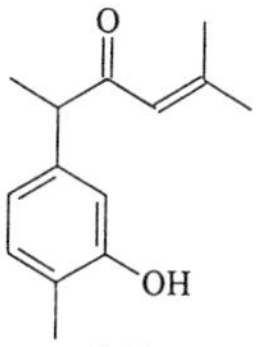

compound 6 (proposed)
(MW=218)

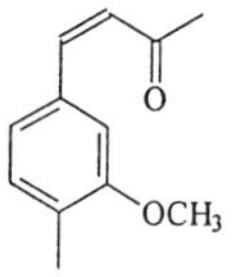

dehydrozingerone
(MW=190)

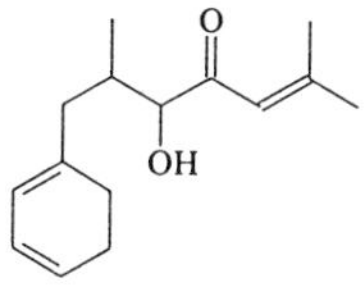

compound 7 (proposed)
(MW=234)

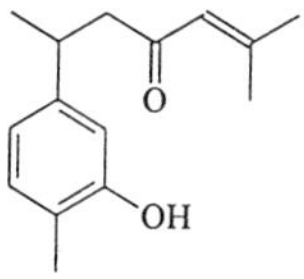

compound 8 (proposed)
(MW=232)

Unidentified
Component

compound 9
(MW=216)

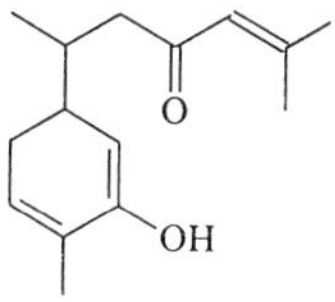

compound 10 (proposed)
(MW=234)

Figure 10. *Continued*

PERCENT is the percent, expressed as a decimal, of turmeric in the turmeric-Chromosorb mixture.

EI-mass spectra were obtained for each component. Figures 7, 8, & 9 show the EI-mass spectra for the three major components detected by direct thermal extraction GC-MS. Table III contains a list of the eight most abundant ions for the minor components in turmeric powder and their relative intensities. Figure 10 lists the structures for the components detected. Some of these identifications are based on correlation with mass spectra in the NIST/EPA/NIH Mass Spectral Library. However, the mass spectra for most of the components could not be found in the library and many of the structures in Figure 10 are proposed structures. For proposed structures, the identifications are based on similarities with components for which there was a good library match and by evaluation of the fragmentation patterns.

Figure 7 shows the EI-mass spectra for *ar*-turmerone, the character impact compound for turmeric. The identification of this component was based on interpretation of the mass spectrum and correlation with previously reported data (*9,10*). Figure 8 shows the mass spectrum of turmerone. Mass spectral data for this component was also reported by Su, *et al* (*9*). They reported a base peak at m/z = 121 with a probe temp equal to 70°C. They also reported the appearance of an ion at m/z = 119 when the probe temperature was increased 150°C which they attributed to the aromatization of the cyclohexadienyl moiety at the high temperatures of the GC-MS interface line. Figure 9 shows the mass spectrum of curlone. This agrees with the data reported by Kisco, *et al* (*12*).

Additional research on the analysis of volatile components in *Curcuma* species can be found in the literature (*10,13,14,15*).

Conclusion

Extraction of natural products yields complex mixtures of volatile, semi-volatile, and nonvolatile components. No one technique for obtaining mass spectral data for these components is generally applicable. Direct thermal extraction GC-MS proved to be a valuable technique for the identification of volatile and semi-volatile components in powdered turmeric.

Acknowledgements

We acknowledge the Center for Advanced Food Technology (CAFT) mass spectrometry facility for providing instrumentation support. CAFT is an initiative of the New Jersey Commission on Science and Technology. This is New Jersey Agricultural Experiment Station (NJAES) publication #D-10570-1-96.

Literature Cited

1) Govindarajan, V. S. *Crit. Rev. in Food Sci. and Nutr.* **1980,** 12, pp. 199-301

2) Masuda, T.; Isobe, J.; Jitoe, A.; and Nakatani, N. *Phytochem.* **1992,** 31, pp. 3645-3647.

3) Toda, S.; Miyase, T.; Arichi, H.; Tanizawa, H.; and Takino, Y. *Chem. Pharm. Bull.* **1985,** 33, pp. 1725-1728.

4) Azuine, M. A. and Bhide, S. V. *Nutr. Cancer* **1992** 17, pp. 77-83.

5) Mukundan, M. A.; Chacko, M. C.; Annapurna, V. V.; and Kirshaswamy, K. *Carcinogenesis* **1993,** 14, pp. 493-496

6) Azuine, M. A.; Kayal, J. J.; and Bhide, S. V. *J. Cancer Res. Clin. Oncol.* **1992,** 118, pp. 447-452

7) Tonnesen, H. H. In *Phenolic Compounds in Food and Their Effects on Health I- Analysis, Occurance, and Chemistry* Editor, Ho, C.-T.; Lee, C. Y.; and Huang, M-T. ACS Symposium Series 507, American Chemical Society: Washington, DC, 1992; pp. 144-153.

8) Hartman, T. G.; Overton, S.; Manura, J.; Baker, C. W.; and Manos, J. N. *Food Technol.* **1991,** 45, pp. 104-105.

9) Su, H. C. F.; Horvat, R.; and Jilani, G. *J. Agric. Food Chem.* **1982,** 30, pp. 290-292.

10) Rao, A. S.; Rajanikanth, B.; and Seshadri, R. *J. Agric. Food Chem.* **1989,** 37, pp. 740-743.

11) Kingston, D. G. I.; Bursey, J. T.; and Bursey, M. M. *Chemical Reviews* **1974,** 74, pp. 215-.

12) Kiso, Y.; Suzuki, Y.; Oshima, Y.; and Hikino, H. *Phytochem.* **1983,** 22, pp. 596-597.

13) Gholap, A. S. and Bandyopadhyay, C. *J. Agric. Food Chem.* **1984,** 32, pp. 57-59.

14) Dung, X. N.; Tuyet, N. T. B.; and Leclercq, P. A. *J. Essent. Oil Res.* **1995,** 7, pp. 261-264.

15) Ky, P. T.; van de Ven, L. J. M.; Leclercq, P. A.; and Dung, N. X. *J. Essent. Oil Res.* **1994,** 6, pp. 213-214.

16) Majlat, P.; Erdos, Z.; and Takacs, J. *J. Chromatogr.* **1974,** 91, pp. 89-103.

17) Khurana, A., and Ho, C-T *J. Liquid Chromatogr.* **1988,** 11, pp. 2295-2304.

18) Hiserodt, R.; Hartman, T.G.; Ho, C.-T.; and Rosen, R.T. *J. Chromatogr.* **1996,** 740, pp. 51-64

Chapter 9

Pungent Flavor Profiles and Components of Spices by Chromatography and Chemiluminescent Nitrogen Detection

E. M. Fujinari

Antek Instruments, Inc., 300 Bammel Westfield Road,
Houston, TX 77090–3508

The pungent characteristic of hot flavors is often due to the presence of a class of nitrogen containing compounds such as capsaicinoids. These components can be analyzed by chromatography with chemiluminescent nitrogen detection (CLND). This sensitive nitrogen-specific detector can simplify complex analyses by eliminating non-nitrogenous components in the sample. This allows the chromatographer to easily focus on the separation of the nitrogen containing components responsible for the "hotness" of spices.

Many kinds of nitrogen containing compounds responsible for pungent or "hot" flavors in spices have been reported. Structures of hot components (I-V) in horseradish oil are shown in Figure 1. The red hot chili peppers contain nitrogenous compounds known as capsaicinoids which are quite similar in structure (Figure 2). Hybrid peppers possess different degrees of hotness, e.g. jalapeno> Indian birds-eye's> Mexican habanero varieties. Piperine (IX) is the hot component in black pepper. Since these analytes contain nitrogen, chromatographic detection using the chemiluminescent nitrogen detector is inherently suitable. Simplified chromatograms are obtained since non-nitrogenous compounds in the samples are transparent to the detector.

Historically, capsaicinoids in foods have been analyzed by organoleptic evaluation (1), colorimetry (2), and UV spectrophotometric methods (3, 4). Chromatographic methods have also been used, including thin layer chromatography (TLC) (5, 6) and gas chromatography (GC) (7-9). Typically, GC methods require a

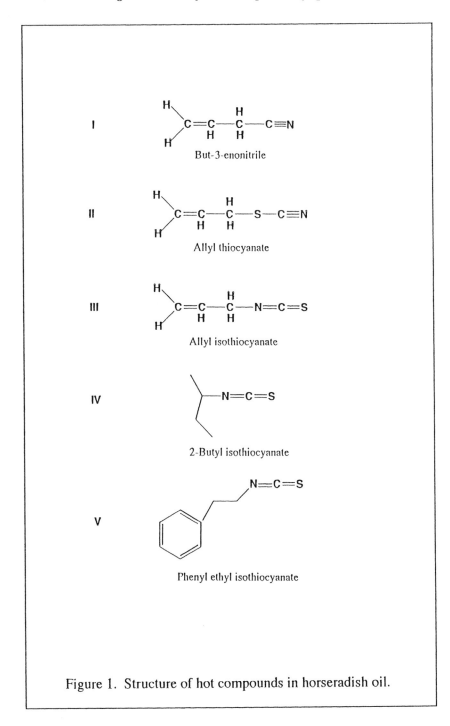

Figure 1. Structure of hot compounds in horseradish oil.

VI

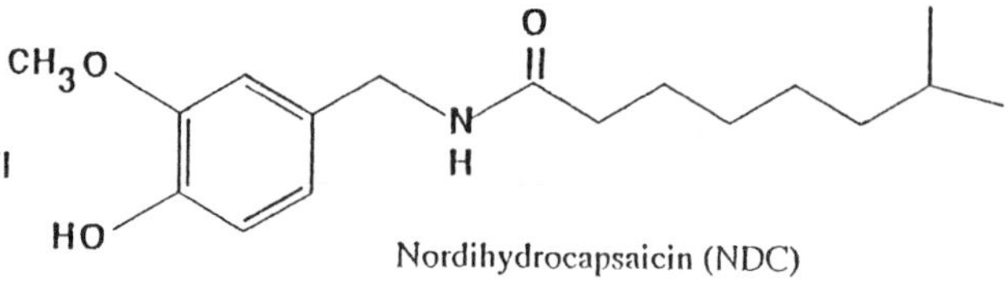

Nordihydrocapsaicin (NDC)

VII

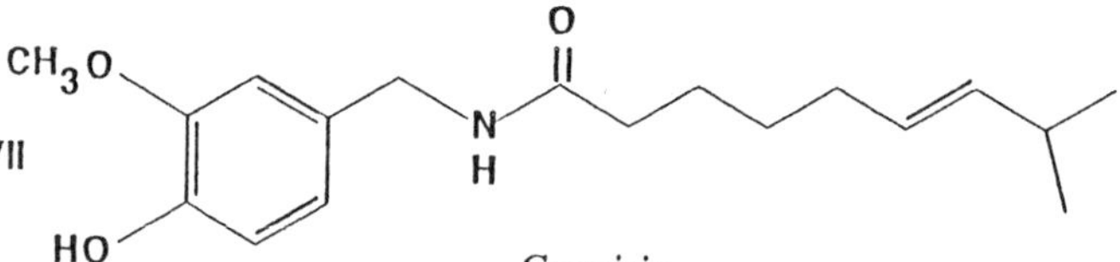

Capsaicin

VIII

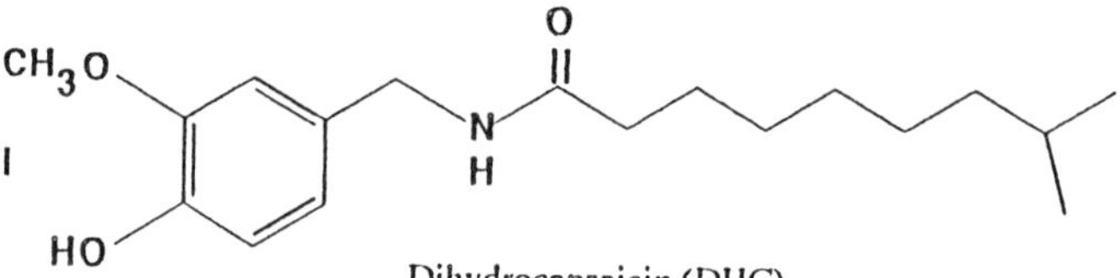

Dihydrocapsaicin (DHC)

IX

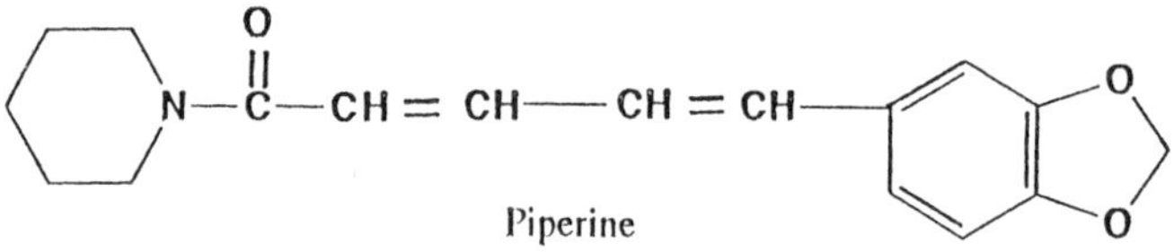

Piperine

Figure 2. Structure of capsaicinoids and piperine.

derivatization step for these compounds prior to analysis in order to make them more volatile. However, capsaicins have been analyzed without derivatization by high performance liquid chromatography (HPLC) with UV detection using normal- and reversed-phase techniques (10-17). Better resolution of capsaicins (17) has been reported using reversed-phase (RP) rather than normal-phase (NP) chromatography. Reversed-phase HPLC separation of piperine followed by UV detection has been reported earlier (14, 18). Using mass spectrometry (MS) as a means for GC detection provides a powerful tool for structural characterization of flavors, e.g. ginger oil (19). However, quantitation of analytes by MS detection in chromatography may be difficult because of the response variation of the detector due to sample and/or solvent induced matrix effects. On the other hand, the CLND response is stable and not affected by complex sample matrices. Quantitation of capsaicin and dihydrocapsaicin in red pepper by HPLC-CLND was previously reported (20). The linear response of the CLND is shown in Figure 3.

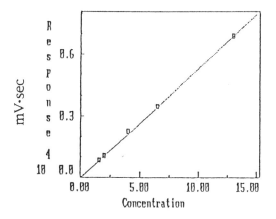

capsaicin (μg)
r=0.99952; m=0.00189; b=-0.00202

Figure 3. HPLC-CLND calibration curve of capsaicin.
Reprinted with permission from E. M. Fujinari, in "Spices, Herbs and Edible Fungi", G. Charalambous (Ed.), 1994, pp 367-379.
with kind permission from Elsevier Science - NL, Sara Burgerhartstraat 25, 1055 KV Amsterdam, The Netherlands.

Benn et. al. ([21]) demonstrated the use of GC-CLND for the detection of nitrogen containing compounds in flavors and essential oils. The advantage of this technique is illustrated in Figures 4a-b using a CP-Sil 5 CB (Chrompack, 25m x 0.53mm ID, 1.0 µm film thickness) column. Figure 4a is the GC-FID profile of a green pepper flavor containing nitrogenous components from the horseradish oil and the three pyrazine compounds. As Figure 4b shows, using GC-CLND, the nitrogen-containing compounds are easily detected without interference from the sample matrix. Peak identification was achieved using the horseradish oil and the pyrazine standards.

Supercritical fluids (such as supercritical CO_2) possess similar viscosities to those of gases, yet their diffusivities are much greater than liquids. These physical properties together provide higher separation efficiencies for SFC, with sharper peaks than for HPLC. Since high molecular weight and thermally labile compounds can be analyzed by this technique, SFC also provides an added advantage over GC. Taylor et. al. ([22]) reported a feasibility study for supercritical fluid chromatography - chemiluminescent nitrogen detection (SFC-CLND) with open tubular columns. SFC-CLND of hot mustard extract is shown in Figure 5. Hot components were identified as allyl isothiocyanate (peak A) and butyl isothiocyanate (peak B) using corresponding analytical standards.

This paper will focus on nitrogen-specific detection for liquid chromatography including HPLC-CLND profiles of chili powder, paprika oleoresin, black pepper, and capsaicins in onion and garlic flavors.

Experimental

Apparatus. High performance liquid chromatographic separations were achieved on a binary gradient microbore HPLC system: primary pump (A) Model 305, secondary pump (B) Model 306, monometric module Model 805, and a dynamic mixer Model 811C from Gilson Inc. (Middleton, WI). Sample injections were achieved with a 20 µL loop on a Model EQ-36 injection valve from Valco Instruments Co. Inc. (Houston, TX). A stainless steel Y-splitter also from Valco was used in order to achieve a post-column split of the mobile phase flow to the CLND. A Supelcosil LC-18S analytical HPLC column was purchased from SUPELCO Inc. (Bellefonte, PA). The Y-splitter was attached to the analytical column by a SLIPFREE connector, available from Keystone Scientific Inc. (Bellefonte, PA). Analyses of nor-dihydrocapsaicin, capsaicin and dihydrocapsaicin in spices as well

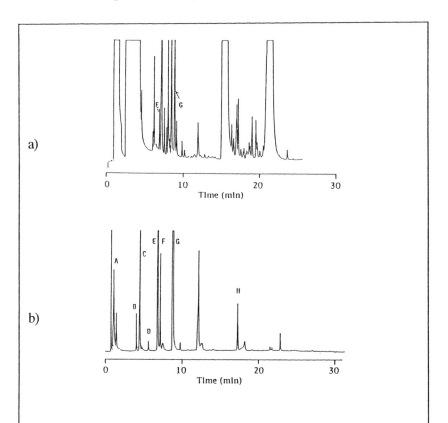

Figure 4. GC profile of green pepper flavor with horseradish
oil and pyrazine mixtures.
a) **FID**: peaks E = 2-methyl-3-methoxypyrazine and
            G = 2-methoxy-3-ethylpyrazine.
b) **CLND**: peaks A = but-3-enonitrile, B = allyl thiocyanate,
C = allyl isothiocyanate, D = 2-butyl isothiocyanate,
E = 2-methyl-3-methoxypyrazine, F = 2-methyl-5-
methoxypyrazine, G = 2-methoxy-3-ethylpyrazine,
and H = phenyl ethyl isothiocyanate. Reprinted with
permission from S. M. Benn, K. Myung, and E. M. Fujinari,
in "Food Flavors, Ingredients and Composition", G.
Charalambous (Ed.), 1993, pp 65-73.
with kind permission from Elsevier Science - NL, Sara
Burgerhartstraat 25, 1055 KV Amsterdam, The Netherlands.

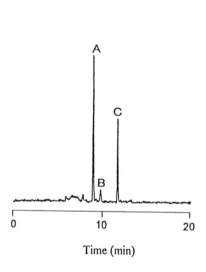

Time (min)

Figure 5. Capillary SFC-CLND profile of hot mustard extract
[0.1 gram of mustard powder extracted in 1 mL water (30%)
and methanol (70%) solution]. Peaks A = allyl isothiocyanate,
B = 2-butyl isothiocyanate, and C = unknown nitrogen
containing compound. Chromatographic conditions:
pressure program from 80 atm (hold 5 min), ramp to 150 atm
at 10 atm/min, then to 200 atm at 15 atm/min; Cyano (20 m x
100 mm ID, 0.25 mm film thickness) column; time split
injection 0.2 sec. Reprinted with permission from H. Shi,
J. T. B. Strode III, E. M. Fujinari and L. T. Taylor,
J Chromatogr. A, 734 (1996) 303.
with kind permission from Elsevier Science - NL, Sara
Burgerhartstraat 25, 1055 KV Amsterdam, The Netherlands.

as piperine in black pepper were accomplished with the nitrogen specific detector, model 7000 HPLC-CLND, from Antek Instruments Inc. (Houston, TX) and Delta chromatography software from Digital Solutions (Margate, Australia) run on an IBM 486 compatible computer.

Reagents and Standards. The natural capsaicin standard mixture was composed of 65% capsaicin and 35% dihydrocapsaicin. Piperine (97%) and citric acid (99+%) were also purchased from Aldrich Chemical Co. (Milwaukee, WI). HPLC grade methanol (99+%) was obtained from Fisher Scientific Co. (Fair Lawn, NJ). Sodium free distilled water was obtained from Ozarka Drinking Water Co. (Houston, TX). All standards and reagents were used without further purification. The HPLC mobile phase was filtered through a Millipore Corp. (Bedford, MA) HV filter with a 0.45 μm pore size. Paprika oleoresin, onion, garlic, black pepper, and chili powders were obtained from commercial sources.

Analytical method. The capsaicin/dihydrocapsaicin and piperine reference standards were prepared as 13.19 mg/mL and 4.16 mg/mL solutions in methanol, respectively. Samples were prepared by separately weighing the following and bringing the volume to 25 mL with methanol: red chili powder (2.5136 g), black pepper powder (2.5154 g), onion powder (1.5126 g), and garlic powder (1.5297 g). Each sample (5 mL) was concentrated to 1 mL final volume with a gentle stream of helium. Samples for the hot garlic and onion flavor profiles were prepared using a (1 + 1 v/v) mixture of the capsaicin reference standard and 100 μL of the concentrated flavors. HPLC-CLND analyses were accomplished using a 15 μL partial filled injection to a 20 μL sample loop into a Supelcosil LC-18S analytical column: 250mm x 4.6mm ID, 5 μm particle size, 100 A pore size. An isocratic mobile phase, methanol/water with 0.1% citric acid at pH 3.0 (65:35 v/v), was utilized with a flow rate of 0.650 mL/min. A Y-splitter was configured post-column and used to deliver a flow of 100 μL/min to the CLND. The CLND conditions were: 1100° C pyrolysis temperature, PMT voltage 780, range x10, and 1 volt detector output.

Results and Discussion

This chemiluminescent nitrogen detector for HPLC was first described in (23). The detection mechanism for nitrogen determination is shown below:

$$R\text{-}N + O_2 \xrightarrow{\quad 1100^{\circ} C \quad} \cdot NO + \text{other oxides}$$

$$\cdot NO + O_3 \longrightarrow \overset{*}{N}O_2 \longrightarrow NO_2 + h\nu$$

Nitrogen containing analytes (R-N) are oxidized in a furnace at high temperatures to nitric oxide (•NO). Chemiluminescence as shown in the second equation is detected by a photomultiplier tube (PMT). The photons detected are proportional to the amount of nitrogen in the analyte(s).

The HPLC-CLND profile of red chili pepper is shown in Figure 6. The capsaicinoids nordihydrocapsaicin (A), capsaicin (B), and dihydrocapsaicin (C) are easily observed without interference from the red chili powder matrix in this chromatogram. This separation was achieved via isocratic reversed-phase chromatography using a mobile phase consisting of MeOH/water with 0.1% citric acid at pH 3.0 (65:35 v/v). A Supelcosil LC-18S (250mm x 4.6mm ID) column with a mobile phase flow rate of 0.65 mL/min was used. A post-column split was used to direct 100 µL/min to the CLND. The three capsaicinoid compounds (A, B, and C) are also very easily detected in onion (Figure 7) and garlic (Figure 8). Several polar nitrogen containing components eluted at the solvent front for the onion, garlic and red chili powder samples. Figure 9 shows the HPLC-CLND profile of black pepper. Paprika oleoresin was extracted by the method reported by Cooper (17) and analyzed by HPLC-CLND (Figure 10) showing the presence of capsaicin (A) and dihydrocapsaicin (B).

Conclusion

Pungent (hot) components can be separated using gas chromato-graphy, supercritical fluid chromatography, and high performance liquid chromatography and detected with the chemiluminescent nitrogen detectors (CLND). These nitrogen-specific detectors provide a means of analyzing nitrogen-containing compounds free of matrix interference.

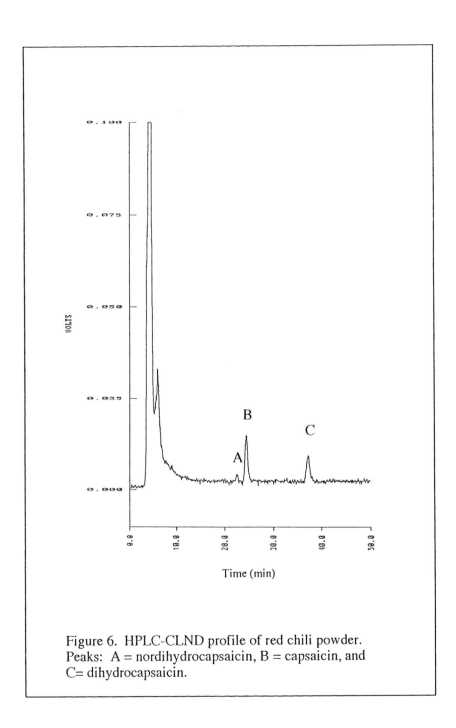

Figure 6. HPLC-CLND profile of red chili powder.
Peaks: A = nordihydrocapsaicin, B = capsaicin, and
C= dihydrocapsaicin.

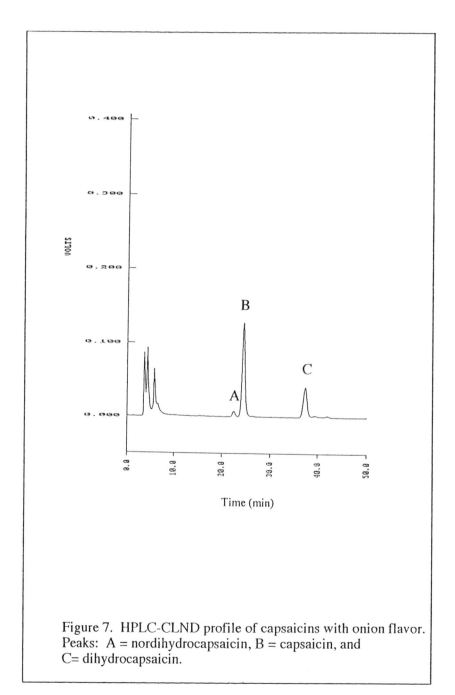

Figure 7. HPLC-CLND profile of capsaicins with onion flavor.
Peaks: A = nordihydrocapsaicin, B = capsaicin, and
C= dihydrocapsaicin.

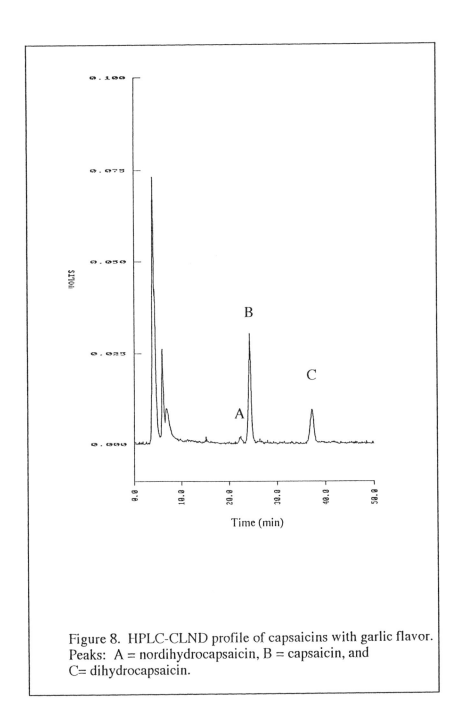

Figure 8. HPLC-CLND profile of capsaicins with garlic flavor.
Peaks: A = nordihydrocapsaicin, B = capsaicin, and
C= dihydrocapsaicin.

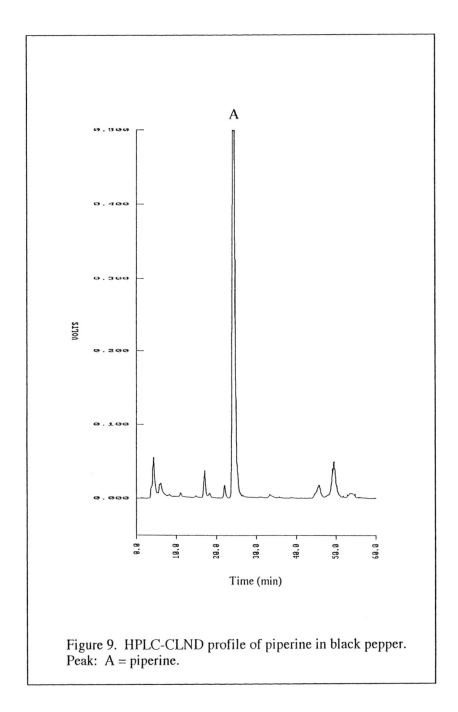

Figure 9. HPLC-CLND profile of piperine in black pepper.
Peak: A = piperine.

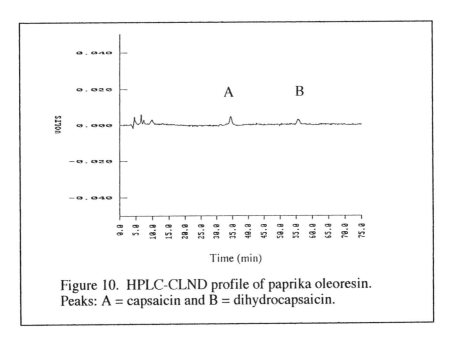

Figure 10. HPLC-CLND profile of paprika oleoresin.
Peaks: A = capsaicin and B = dihydrocapsaicin.

Literature Cited

1. Govindarajan, V.S.; Narasimhan, S.; Dhanara, S.J. J. Food Sci. Technol. 1977, 14 , 28-34.
2. Van Fedor, K.; Unters, Z. Lebensm 1931, 61, 94-100.
3. DiCecco, J.J. J. Assoc. Off. Anal. Chem. 1979, 62, 998-1000.
4. Trejo, G.; Wila, A. J. Food Sci. 1973, 38, 342-344.
5. Spanjar, P.; Blazovich, M. Analyst 1969, 94, 1084-1089.
6. Andre, L.; Mile, L. Acta Alimen. 1975, 4, 113-121.
7. Hartman, K.T. J. Food Sci. 1970, 35, 543-547.
8. DiCecco, J. J. Assoc. Off. Anal. Chem. 1976, 59, 1-3.
9. Todd, P.H.; Bensinger, M.G.; Biftu, T. J. Food Sci. 1977, 42, 660-668.
10. Woodbury, J.E. J. Assoc. Off. Anal. Chem., 1980, 63, 556-558.
11. Saria, A.; Lembeck, F.; Skofitsch, G. J. Chromatogr. 1981, 208, 41-46.
12. Hoffman, P.G.; Lego, M.C.; Galetto, W.G. J. Agric. Food Chem.. 1983, 31, 1326-1330.
13. Law, M.W. J. Assoc. Off. Anal. Chem. 1983, 66, 1304-1306.
14. Weaver, K.M.; Luker, R.G; Neale, M. E. J. Chromatogr. 1984, 301, 288-291.
15. Kawada, T.; Watanabe, T.; Katsura, K.; Takami, H.; Iwai, K. J. Chromatogr. 1985, 329, 99-105.

16. Weaver, K.M.; Awde, D.B. J. Chromatogr. 1986, 367, 438-442.
17. Cooper, T.H.; Guzinski, J.A.; Fisher, C. J. Agric. Food Chem. 1991, 39 2253-2256.
18. Wood, A.B.; Barrow, M.L.; James, D.J. Flavour Fragr. J. 1988, 3, 55-64.
19. Vernin, G.; Parakanyi, C. In *Spices, Herbs, and Edible Fungi*; Charalambous, G., Ed.; Developments in Food Science Series 34; Elsevier Science Publishers, Amsterdam, The Netherlands, 1994, 34 579-594.
20. Fujinari, E.M. In *Spices, Herbs, and Edible Fungi*; Charalambous, G., Ed.; Developments in Food Science Series 34; Elsevier Science Publishers, Amsterdam, The Netherlands, 1994, 34 367-379.
21. Benn, S. M.; Myung, K.; Fujinari, E. M. In *Food Flavors, Ingredients and Composition*; Charalambous, G., Ed.; Developments in Food Science Series 32; Elsevier Science Publishers, Amsterdam, The Netherlands, 1993, 32 65-73.
22. Shi, H.; Strode, J.T.B.III; Fujinari, E.M.; Taylor, L.T. J. Chromatogr.A, 734 (1996) 303-310.
23. Fujinari E. M.; Courthaudon L. O. J. Chromatogr. 1992, 592, 209-214.

Chapter 10

Supercritical Fluid Extraction of *Allium* Species

Elizabeth M. Calvey[1] and Eric Block[2]

[1]Center for Food Safety and Applied Nutrition, U.S. Food and Drug Administration, 200 C Street Southwest, Washington, DC 20204
[2]Department of Chemistry, State University of New York—Albany, Albany, NY 12222

Supercritical fluid technologies are viable alternatives for the extraction and analysis of natural products because of the heightened awareness of the cost and safety hazards associated with the use and disposal of conventional organic solvents. Supercritical CO_2 (SC-CO_2) is of particular interest to the food industry. Because of its low critical temperature, SC-CO_2 extraction can provide an accurate representation of the taste, color and odor of naturally occurring materials found in spices, flavors and foods. The application of supercritical fluid extraction technology to the analysis and/or production of spice and flavor extracts is discussed. The use of the technology in the extraction of organosulfur compounds found in *Allium* species is emphasized.

Applications of supercritical fluid extraction (SFE) technologies are being investigated extensively by the food industry (1-9). The extensive patent activity related to food processing applications since the early 1970s is evidence of interest in SFE. Over 50 patents have been issued or applied for in the US or elsewhere. Table I lists some of the patents related to the flavor and spice industry (10-23). Supercritical CO_2 (SC-CO_2) with its low critical temperature (31.3 °C) is of particular interest to the food industry because many of the components in food matrices react or degrade at elevated temperatures. The advantages of SFE for the food industry include potentially higher yields, better quality products and the use of a nonflammable, nontoxic solvent. Current Food and Drug Administration regulations [21 *Code of Federal Regulations* 184.1240(c)] list CO_2 as "generally recognized as safe" when used as a direct human food ingredient. For this reason, CO_2 can be used in food with virtually no limitations other than current good manufacturing practice. In contrast, the use of traditional organic solvents requires the removal of residual solvent to permitted levels. This removal usually requires some distillation, which can cause off-flavors due to decomposition of components at the elevated temperatures used. The chance of off-flavors resulting from residual solvents is eliminated when extracting with SC-CO_2 because of the ease of removing CO_2 from the food matrix. Other motives for the investigation of SFs include the potential for new product development and compliance with more stringent pollution control regulations that increase the cost of waste disposal of traditional solvents (24).

Although SC-CO_2 can solubilize non-polar to moderately polar compounds, the use of entrainers or modifiers (e.g., ethanol) enhances the solvating properties of the

TABLE I. Representative Patents Related to SFE of Flavors and Spices

Patent #	Date	Title	Reference
US 3 477 856	11/11/69	Process for the Extraction of Flavors	10
US 4 123 559	10/31/78	Process for the Production of Spice Extracts	10,11
DE 3 133 032	3/3/83	Apparatus and Methods for Extraction	12
US 4 400 398	8/23/83	Method for Obtaining Aromatics and Dyestuffs from Bell Peppers	10
US 4 470 927	9/11/84	Method of Extracting the Flavoring Substances from the Vanilla Capsule	10,13
US 4 474 994	10/2/84	Purification of Vanillin	10,14
US 4 490 398	12/25/84	Process for the Preparation of Spice Extracts	10,15
JP 84 232 064	12/26/84	Manufacture of Flavoring Substances	16
EP Appl. 206 738/ US Appl. 746 607	12/30/86 6/19/85	Process for the Production of Citrus Flavor and Aroma Compositions	17
EP Appl. 154 258	12/11/85	Flavoring Extracts	18
JP 63 87 977	4/19/88	Isolation of Spices, Peroxides and Other Useful Components from Cruciferous Plants	19
JP 01 117 761	5/10/89	Extraction of Flavor Components of Alcoholic Beverages	20
JP 02 135 069	5/23/90	Manufacture of Spice Extracts	21
JP 02 235 997	9/18/90	Extraction of Flavoring from Seaweeds	22
US 5 120 558	6/9/92	Process for the Supercritical Extraction and Fractionation of Spices	23

fluid. Whereas a neat compound may be soluble in $SC-CO_2$, it may not be extractable from its matrix without the addition of an entrainer. This phenomenon is demonstrated in the decaffeination of coffee (10); neat caffeine is soluble in dry $SC-CO_2$, but moist $SC-CO_2$ or moist coffee is necessary for the extraction of caffeine from coffee beans. This same phenomenon occurs with decaffeination by traditional organic solvents. Investigators have hypothesized that water frees the "chemically bound" caffeine in the coffee matrix.

An SFE system contains five basic components: pump(s), extraction vessel, temperature controls, pressure controls and separator(s). For processing, a variety of recovery strategies is viable: a) change of temperature; b) change of pressure; and/or c) use of a suitable adsorbent material. The complexity of the processing SFE system depends on the desired application and the mode of product recovery. Rizvi et al. (8) divided food-related SFE applications into three basic categories: a) total extraction, b) deodorization, and c) fractionation. Total extraction is the removal of a component or group of related compounds from an insoluble matrix. This type of process is exemplified by the extraction of vegetable oils with $SC-CO_2$. Deodorization relates to operating the extraction system at less than the maximum density of the solvent. The extraction conditions are usually held constant while the components of interest, generally the more soluble ones, are preferentially removed from the matrix. This type of application is appropriate for the removal of objectionable aromatics or the extraction of desirable odor components such as from spices. Whereas SFE in general fractionates a matrix, Rizvi et al. (8) use the term fractionation to describe the separation of coextracted components from each other. They also use the term to describe concentration of components either as the extractant or in the residual material. In sample preparation, the SFE system can be directly attached to a chromatographic system and used as an on-line injection technique. If SFE is performed off-line, the resulting extractant is introduced into a chromatographic system via conventional injection techniques. Multiple examples of on-line and off-line analytical scale SFE related to food products can be found in the literature (9,25-44).

SFE of Spices and Flavors

Although many compounds have been extracted by using $SC-CO_2$, the majority of the work related to food products can be divided into three broad categories: flavors/spices, herbicides/pesticides and lipids. Table II lists the extraction conditions which use $SC-CO_2$ for a variety of spice and flavor applications. The loss of the volatile components desired in spice extracts is reduced because CO_2 is a gas at room temperature and the higher temperatures required in distillation are not needed. Hubert and Vitzthum (47) indicated that the best spice extracts have all the organoleptic factors of the whole spice even after dilution of the extract. They investigated black pepper, nutmeg and chilies and indicated that the products obtained from the $SC-CO_2$ extractions were organoleptically similar to the commercially obtained extracts. In 1981 Caragay (48) reviewed the extraction conditions for cloves, cinnamon and vanilla pods. Stahl and Gerard (45) studied the solubility and fractionation of essential oils in $SC-CO_2$. They were able to obtain quantitative recovery of these volatile oils free of undesirable substances without fractionation of the essential oils themselves. Once the essential oil components were extracted, further fractionation into certain substance groups (i.e., monoterpene hydrocarbons, sesquiterpene hydrocarbons, oxygen containing monoterpenes and oxygen containing sesquiterpenes) was possible. Several laboratories have employed off-line extraction techniques to fractionate the flavor components of ginger, pimento berries, apple essence (29) and lemon peel (30). By using gas chromatography (GC) as the means of analysis, these laboratories were able to show some fractionation of the flavor components as a function of the extraction density. Other laboratories have investigated the use of SFE in flavor analysis by direct coupling of the extractor to a chromatographic technique such as GC (25,26,33,38) and SFC (27,34).

TABLE II. Representative Applications of SFE Related to Flavors and Spices

Commodity	Conditions	Reference
Rosemary	45 °C; 300 atm; 10 min; cryotrapping -65 or -10 °C	25
Eucalyptus leaves, lime peel, lemon peel, basil	45 °C; 300 atm; 10 min; cryogenic trapping -50 to 30 °C	26
Cold-pressed grapefruit oil	70 °C; 0.1767-0.8579 g/mL; 12 min; cryotrapping -65 or -10 °C	27
Garlic and onion	35 °C; 240 atm; 30 min; cryotrapping 0-1 °C	28
Ginger, pimento berries	50 °C; 1500-5000 psi	29
Lemon peel oil	30-58 °C; 90-250 kg/cm^2	30
Mexican spices (*Origanum vulgare* and *Pimpinella anisum*)	55 °C; 167 bars; separation vessel: 55 °C; 26.7 bar.	31
Chamomile essential oil	40 °C; 90 bar; separation vessels in series: 0 °C, 90 bar; -5 °C, 30 bar.	32
Cloves	40-140 °C; 82-400 atm; 4.5 h	33
Turmeric	60 °C; 250-280 bar; 15-20% MeOH	34
Shiitake mushrooms	40 °C; 3000 psi; fractionation through several traps	35
Hops essential oil, bitter principles	50 °C; 0.05-0.1 g/mL; 15-30 min	36
Lavender essential oil, waxes	48 °C; 90 bar; separation vessels in series: -10 °C, 80 bar; 0 °C, 25 bar.	37
Caraway fruits	50 °C; 9.7 MPa	38
Ginger	40 °C; 0.85 g/mL; 3 min static, 30 min dynamic; 1 mL/min	39
Essential oil components (limonene, carvone, anethole eugenol, caryophyllene, valeranone)	40-120 °C; 20-120 bar	45
Orange juice (treatment used to deactivate pectinesterase)	35-60 °C; 7-34 MPa; 15-180 min	46

SFE of *Allium* Species

The natural flavors from garlic (*Allium sativum*), onion (*A. cepa*), and other *Allium* species, like those from many other common vegetables and fruits, are not present as such in the intact plants but are formed by enzymatic processes when the plants are chewed or cut (49). Additional flavors, also considered natural, are formed during cooking as a result of the thermal breakdown of the initial enzymatically produced flavorants in either an aqueous on nonaqueous (e.g., cooking oil) medium. If the breakdown products are unstable, other compounds can be formed, which can contribute to the aroma and taste of the food.

Supercritical fluid extraction of *Allium* species with CO_2 provides an effective and environmentally friendly alternative to traditional organic extractions. Miles and Quimby (50) extracted garlic products by using SC-CO_2 under mild conditions. They analyzed the extracts by GC with atomic emission detection. Sinha (51) extracted onions with SC-CO_2 and analyzed the extracts by GC-mass spectrometry (MS). The use of traditional GC methodologies in the analyses of these SF extracts by the above research groups precluded the identification of those compounds primarily responsible for the characteristic flavors of freshly cut members of the genus *Allium* (52). We have previously shown that the chromatographic profiles of extracts of *Allium* species (garlic and onion) obtained with SC-CO_2 at the low temperature of 35°C were similar to the chromatographic profiles of corresponding organic solvent extracts (28). The flavor qualities of the SF extracts were judged to be comparable to those of fresh garlic and onion. Our experience with SFE of garlic contrasts with that of Wagner and Breu (53). They reported the complete decomposition of garlic compounds, such as allicin (S-2-propenyl 2-propenethiosulfinate, AllS(O)SAll), after garlic juice was stored at room temperature for 3 h and then extracted with CO_2. It is difficult to evaluate their work because they did not report the SFE conditions employed. We observe very little change in the liquid chromatographic profile of SF extracts of aqueous homogenates of garlic that were extracted at 35°C following storage for 10 min or 2 h at room temperature (27.5°C) (Figure 1). Our observations agree with those of Lawson (54), who has shown that the half-life of allicin at room temperature is 4 days in water.

Supercritical fluid extraction of the major garlic flavorant, allicin, is 25% more efficient than the best procedure which uses organic solvents if the SFE is done at or below 35°C. Liquid chromatography (LC)-MS under thermospray conditions confirmed the identity of allicin from the garlic (28). Analyses of SF extracts of commercial garlic products which have been reconstituted by addition of water showed LC profiles similar to those seen with fresh garlic. Allicin and related thiosulfinates (RS(O)SR') have been identified in SF extracts of garlic (Figure 2A), freeze-dried ramp (*A. tricoccum*) and frozen ramp (Figure 2B). The ratio of the thiosulfinates found are significantly different in the garlic and ramp. The major constituent in the garlic extract is allicin, peak (pk) 4. The major constituents in the frozen ramp extract are allyl methyl thiosulfinates (AllS(O)SMe, pks 2/3) with a significant contribution of the dimethyl thiosulfinate (MeS(O)SMe, pk 1). The 1-propenyl isomers of allicin are also present in both the garlic and ramp extracts. The chromatographic profile of the freeze-dried ramp extract more closely resembles the garlic extract than the extract from the frozen whole ramp. Ajoene (AllS(O)CH$_2$CH=CHSSAll), a major component found in oil-macerated garlic products, has been found in small quantities in SF extracts of garlic and (tentatively) in those from ramp. The identity of these compounds has been verified by LC-MS employing atmospheric pressure chemical ionization (unpublished data).

A similar study employing onion juice showed SFE to be ca. 69% as efficient as conventional organic solvents for extracting the major organosulfur compounds (28). The diminished efficiency of SFE for onion compared with garlic may be due to

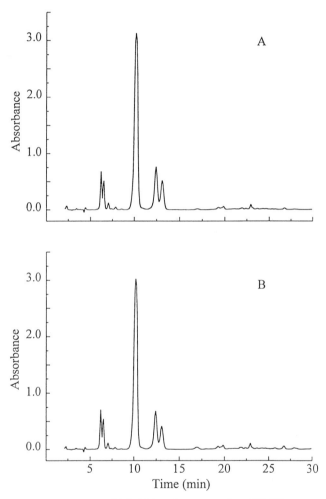

Figure 1. LC chromatograms (UV, 254 nm) representing: (A) SF extract of a garlic homogenate stored at room temperature for 10 min and (B) SF extract of a garlic homogenate stored at room temperature for 2 h.

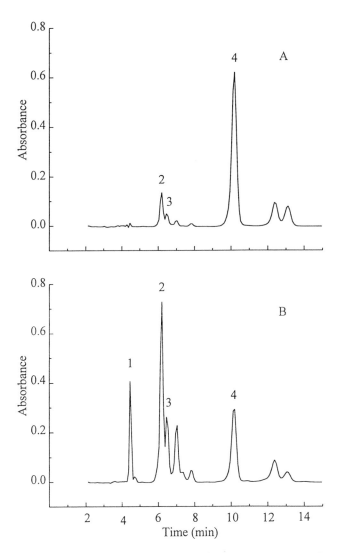

Figure 2. LC chromatograms (UV, 254 nm) of SF extracts: (A) garlic and (B) frozen fresh ramps. Peak identification: (1) MeS(O)SMe; (2/3) AllS(O)SMe; and (4) allicin.

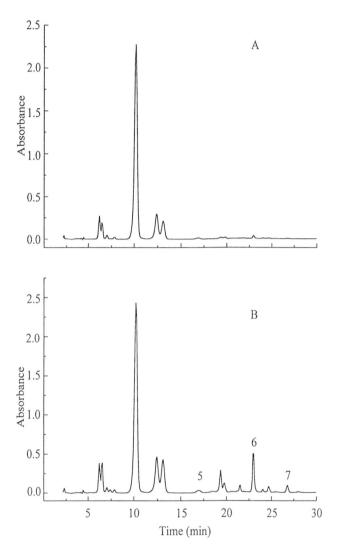

Figure 3. LC chromatograms (UV, 210 nm) of garlic SF extracts: (A) 35 °C, 3 min static period at 240 atm. (B) 50 °C, 28 min static period at 240 atm. Peak identification: (5) ajoene(s); (6) 1,3-vinyldithiin; and (7) diallyl trisulfide.

the 100-fold decrease in levels of flavorants in the former and/or our failure to detect non-volatile flavorants present in the SF extracts of onions. The profile of the flavorants in SF extracts of onion, as determined using GC-MS, was very similar to that previously reported for ether extracts of onion (55). Analysis of SF extracts of onion by LC showed many more components than seen by GC-MS but peak overlap makes component identification difficult. In brief, analysis of an SF extract rapidly prepared from 10-15 g of fresh yellow onions and immediately analyzed showed the presence of various thiosulfinates, propanethial S-oxide (LF), bissulfine(s), zwiebelanes, significant quantities of a series of cepaenes (RS(O)CHEtSSR') and a few related compounds (56,57). Because many of these compounds were non-volatile, GC-MS may have significantly underestimated the efficiency of the SFE procedure. Moreover many of these compounds had weak UV absorption so that monitoring LC analyses by UV detection at 254 nm was problematic. Instead, the identification of compounds was achieved by LC/MS techniques that resolved overlapping compounds of different masses (58). This analysis employed synthetic standards that had been authenticated by other spectroscopic methods.

The Effect of SFE Conditions on the Distribution of Organosulfur Compounds Found in Garlic

Yields of biologically active rearrangement products (ajoenes, dithiins and alkyl sulfides) derived from allicin vary with the type of solvent used. Iberl et al. (59) reported on the transformation products from crushed garlic that had been stored in ethanol for 9 days (216 h). Approximately 10% of the initial allicin was converted to the ajoenes and 8% of the allicin remained; the major product formed from the initial thiosulfinates present was the diallyl trisulfide (>70%). This pattern contrasted with Iberl et al's results for crushed garlic stored in oil. In this case, no allicin remained after 24 h, and the ajoenes, dithiins and diallyl disulfide constituted ca. 25, 60 and 10% of the transformation products, respectively. At 48 and 216 h slight changes in these ratios of the transformation products were found in the crushed garlic stored in oil.

We have continued our SFE studies of *Allium* species by investigating the effects of different SFE conditions (i.e., temperature, static extraction time and modifiers) on the distribution of organosulfur compounds in the resulting extracts. These extracts were analyzed by reversed-phase LC. Initially, we investigated the effect of temperature and the duration of static extraction time on the distribution of the organosulfur compounds in garlic SF extracts by comparing chromatographic profiles. An SF extract obtained at 35°C with a 5 min static extraction time was used as the reference. Our preliminary results showed that an SF extract obtained at 50°C with a 30 min static period produced the most dramatic changes in the chromatographic profiles. These changes were observed after 15 min (Figure 3). By comparing LC retention times and on-line UV spectra with synthesized standards (Figure 4), two of the peaks in this region can be assigned as ajoene(s) (pk 5) and diallyl trisulfide (pk 7). The most intense peak in this region can be tentatively identified as 1,3-vinyldithiin (pk 6) by comparing its UV spectra with that found in the literature (59). The other peaks in this region have UV spectra characteristic of alkylsulfides.

Experimental

Extraction. Allium extracts were obtained by using a PrepMaster and Accutrap (Suprex, Inc., Pittsburgh, PA). Whole cloves of garlic (2-4 g) or ramp bulbs (2-5 g) were homogenized in 10 mL/g of water at ambient temperature with a Tissumizer (Tekmar, Cincinnati, OH). Freeze dried ramp (0.5 g) was homogenized with 20 mL of water. The solutions were allowed to stand at room temperature for 5-10 min to ensure complete enzymatic conversion to the thiosulfinates and related compounds. Hydromatrix (Varian, Palo Alto, CA) was mixed with 15 mL of homogenate and

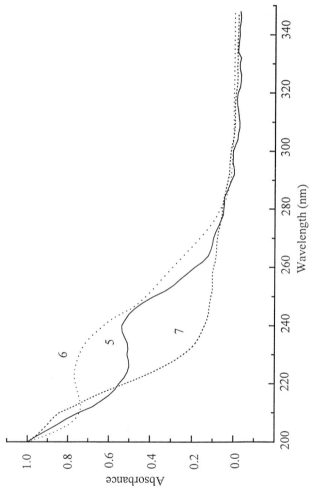

Figure 4. UV spectra of components found in garlic SF extract (50 °C, 28 min static period at 240 atm). Each spectrum was independently normalized. Peak identification: (5) ajoene(s); (6) 1,3-vinyldithiin; and (7) diallyl trisulfide.

placed in a 50 mL extraction vessel. Unless otherwise indicated, the homogenates were extracted at 35°C and 240 atm at a flow rate of 2 mL/min until 60 g of CO_2 (SFE/SFC grade, Air Products, Allentown, PA) were consumed. A static extraction time of 5 min was used to ensure thermal equilibrium. The effluent was trapped on glass beads at 1°C and desorbed with 1 mL of methanol, methanol/water (50/50), or ethanol. The ramp extracts were analyzed directly, whereas the garlic extracts were diluted to 5 mL with water.

LC Analyses. LC analyses were performed by using a Waters 600 pump equipped with a Rheodyne Model 7725i manual injector (50 µL loop) and a Waters Model 991 photodiode array detection system. Chromatographic separation was achieved on a YMC-Pack PolymerC18 (YMC Inc. Wilmington, NC) column with a OPTI-SOLV mini filter (Optimize Technologies Inc., Portland, OR). The linear gradient was 60% H_2O (solvent A)/40% acetonitrile (solvent B) which was held for 10 min, then increased by 8% acetonitrile/min to 20% H_2O/80% acetonitrile, and held for 15 min. The flow rate was 0.8 mL/min.

Acknowledgments

We gratefully acknowledge support from the National Science Foundation (E.B.), the NRI Competitive Grants Program/U.S. Department of Agriculture (Award No. 92-37500-8060), and McCormick & Company, Inc. (E.B.). We thank R. Epply for walking the banks of the Potomac in search of wild ramp and P. Whanger for providing the freeze dried ramp.

Literature Cited

1. Grimmett, C. *Chemistry and Industry*, **1981**, 359.
2. King, J.W.; Johnson, J.H.; Friedrich, J.P. *J. Agric. Food Chem.* **1989**, *37*, 951.
3. Hoyer, G.C. *ChemTech* **1985**, 440.
4. Bartle, K.D.; Clifford, A.A. *Spec. Publ. - R. Soc. Chem.* **1994**, *160*, 1.
5. King, J.W., *INFORM* **1993**, *4*, 1089.
6. Eckert, C.A.; Van Alsten, J.G.; Stoicos, T. *Environ. Sci. Technol.* **1986**, *20*, 319.
7. Dziezak, J.D. *Food Technology* **1986**, *June*, 66.
8. Rizvi, S.S.H.; Daniel, J.A.; Benado, A.L.; Zollweg, J.A. *Food Technology* **1986**, 57.
9. Hawthorne, S.B. *Anal. Chem.* **1990**, *62*, 633A.
10. McHugh, M.A.; Krukonis, V.J., Eds. *Supercritical Fluid Extraction Principles and Practice*, Butterworths, Boston, 1986.
11. Vitzthum, O.; Hubert, P. US Patent 4 123 559, 1978; *Chem. Abstr.*. **1973**, *79*, 30678a.
12. Strauss, K. Ger. Offen. DE Patent 3 133 032, 1983; *Chem. Abstr.*. **1983**, *99*, 7546k.
13. Schütz, E.; Vollbrecht, H.-R.; Sandner, K.; Sand, T.; Mühlnickel, P. US Patent 4 470 927, 1984; *Chem. Abstr.* **1983**, *99*, 4363n.
14. Makin, E.C. US Patent 4 474 994, 1984; *Chem. Abstr.* **1984**, *101*, 210742y.
15. Behr, N.; Van der Mei, H.; Sirtl, W.; Schnegelberger, H.; Von Ettinghausen, O. US Patent 4 490 398, 1984; *Chem. Abstr.* **1985**, *102*, 111886n.
16. Japan Oxygen Co., Ltd. Jpn. Kokai Tokkyo Koho JP Patent 59 232 064, 1984; *Chem. Abstr.* **1985**, *102*, 202893d.
17. Japikse, C.H.; Van Brockin, L.P.; Hembree, J.A.; Kitts, R.R.; Meece, D.R. Eur. Pat. Appl. EP 206 738, 1986; *Chem. Abstr.* **1987**, *107*, 6048w.
18. Calame, J.P.; Steiner, R. Eur. Pat. Appl. EP 154 258, 1985; *Chem. Abstr.* **1986**, *104*, 33288u.
19. Kobayashi, T.; Taniguchi, M. Jpn. Kokai Tokkyo Koho JP Patent 63 87 977, 1988; *Chem. Abstr.* **1988**, *109*, 207527s.

20. Sugyama, K.; Tomohiro, Y. Jpn. Kokai Tokkyo Koho JP Patent 01 117 761, 1989; *Chem. Abstr.* **1989**, *111*, 76540g.
21. Kazuyuki, Y.; Kobayashi, M. Jpn. Kokai Tokkyo Koho JP Patent 02 135 069, 1990; *Chem. Abstr.* **1990**, *113*, 96385e.
22. Kobayashi, M.; Shiraishi, S.; Matsukura, K.; Yamashita, K. Jpn. Kokai Tokkyo Koho JP Patent 02 235 997, 1990; *Chem. Abstr.* **1991**, *114*, 60706s.
23. Nguyen, U.; Evans, D.A.; Berger, D.J.; Calderon, J.A. US patent 5 120 558, 1992.
24. Coenen, H.; Kriegel, E. *Ger. Chem. Eng.* **1984**, *7*, 335.
25. Hawthorne, S.B.; Krieger, M.S.; Miller, D.J. *Anal. Chem.* **1988**, *60*, 472.
26. Hawthorne, S.B.; Miller, D.J.; Krieger, M.S. *J. Chromatogr. Sci.* **1989**, *27*, 347.
27. Anderson, M.R.; Swanson, J.T.; Porter, N.L.; Richter, B.E. *J. Chromatogr. Sci.* 1989, 27, 371.
28. Calvey, E.M.; Matusik, J.E.; White, K.D.; Betz, J.M.; Block, E.; Littlejohn, M.H.; Naganathan, S.; Putman, D.J. *J. Agric. Food Chem.* **1994**, *42*, 1335.
29. Krukonis, V.J. *ACS Symp. Ser.* **1985**, *289*, 154.
30. Sugiyama, K.; Saito, M. *J. Chromatogr.* **1988**, *442*, 121.
31. Ondarza, M. ; Sanchez, A. *Chromatographia* **1990**, *30*, 16.
32. Reverchon, E.; Senatore, F. *J. Agric. Food Chem.* **1994**, *42*, 154.
33. Huston, C.K.; Ji, H. *J. Agric. Food Chem.* **1991**, *39*, 1229.
34. Sanagi, M.M.; Ahmad, U.K.; Smith, R.M. *J. Chromatogr. Sci.* **1993**, *31*, 20.
35. Charpentier, B.A.; Sevenants, M.R.; Sanders, R.A. in *The Shelf Life of Foods and Beverages*, Charalambous, G. (Ed.), Elsevier, Amsterdam, **1986**, 413.
36. Verschuere, M.; Sandra, P.; David, F. *J. Chromatogr. Sci.* **1992**, *30*, 388.
37. Reverchon, E.; Della Porta, G.; Senatore, F. *J. Agric. Food Chem.* **1995**, *43*, 1654.
38. Kallio, H.; Kerrola, K.; Alhonmaki, P. *J. Agric. Food Chem.* **1994**, *42*, 2478.
39. Bartley, J.P. *J. Sci. Food Agric.* **1995**, *68*, 215.
40. Taylor, S.L.; King, J.W.; Snyder, J.M. *J. Microcolumn Sep.* **1994**, *6*, 467-473.
41. Blanch, G.P.; Ibanez, E.; Herraiz, M.; Reglero, G. *Anal. Chem.* **1994**, *66*, 888-892.
42. Nam, K.S.; King, J.W. *HRC* **1994**, *17*, 577-582.
43. Lembke, P.; Bornet, J.; Englehardt, H. *J. Agric. Food Chem.* **1995**, *43*, 38-55.
44. Blanch, G.P.; Reglero, G.; Herraiz, M. *J. Agric. Food Chem.* **1995**, *43*, 1251-1258.
45. Stahl, E.; Gerard, D. *Perfumer & Flavorist.* **1985**, *10*, 29.
46. Arreola, A.G.; Balaban, M.O.; Marshall, M.R.; Peplow, A.J.; Wei, C.I.; Cornell, J.A. *J. Food Sci.* **1991**, *56*, 1030.
47. Hubert, P.; Vitzthum, O.G. *Angew. Chem. Int. Ed. Engl.* **1978**, *17*, 710.
48. Caragay, A.B. *Perfumer & Flavorist.* **1981**, *6*, 43.
49. Virtanen, A.I. *Phytochem.* **1965**, *4*, 207.
50. Miles, W.S.; Quimby, B.D. *Am. Lab.* **1990**, *July*, 28F.
51. Sinha, N.K.; Guyer, D.E.; Gage, D.A.; Lira, C.T. *J. Agric. Food Chem.* **1992**, *40*, 842.
52. Block, E. *J. Agric. Food Chem.* **1993**, *41*, 692.
53. Wagner, H.; Breu, W. *Dtsch. Apoth. Ztg.* **1989**, *129*, 21.
54. Lawson, L.D. in *Human Medicinal Agents from Plants*, Kinghorn, A.D.; Balandrin, M.F. (Eds.) ACS Symposium Series No. 534, American Chemical Society, Washington, D.C. **1993**, 306-330.
55. Block, E.M.; Putman, D.; Zhao, S.-H. *J. Agric. Food Chem.* **1992**, *40*, 2431.
56. Block, E. *Angew. Chem. Int. Ed. Engl.* **1992**, *31*, 1135-1178.
57. Block, E.; Calvey, E.M. in *Sulfur Compounds in Foods*, Mussinan, C.J.; Keelan, M.E. Eds. ACS Symposium Series 564, ACS, Washington, D.C. **1994**, 63-79.
58. Matusik, J.E.; White, K.D.; Block, E.; Calvey, E.M. "Supercritical Fluid Extraction and Atmospheric Pressure Chemical Ionization MS/MS of Cepaenes." presented at 44th ASMS Annual Meeting. Portland OR. May **1996**.
59. Iberl, B.; Winkler, G.; Knobloch, K. *Planta Med.* **1990**, *56*, 202.

Chapter 11

Determination of Glucosinolates in Mustard by High-Performance Liquid Chromatography—Electrospray Mass Spectrometry

Carol L. Zrybko[1] and Robert T. Rosen[2]

[1]Nabisco, Inc., 200 DeForest Avenue, East Hanover, NJ 07936
[2]Center for Advanced Food Technology, Cook College, Rutgers, The State University of New Jersey, New Brunswick, NJ 08903–0231

A method was developed to determine glucosinolates in mustard seeds by reverse phase HPLC using volatile ion-pairing reagents followed by UV detection at 235 nm. Two external standards, phenethylglucosinolate and sinigrin were used to quantify results. The identities of individual mustard glucosinolates were confirmed by negative ion electrospray mass spectrometry as was LC peak purity. This LC/MS method may be used to identify species which are not commercially available as pure standards since mass spectrometry can be used to check for all known glucosinolate anions.

Three mustard types were chosen from the *Brassica* species: yellow (*Brassica hirta*), brown, and oriental (*Brassica juncea*). Eleven mustard samples representing harvest areas of southwestern Canada were analyzed in triplicate for glucosinolate content. Percent coefficient of variation between triplicate samples of the same batch was often less than 10%.

In mustard and other members of the *Brassica* species of the Cruciferea family of vegetables, the important nonvolatile precursors are the hydrophilic glucosinolates. The structure of glucosinolates, as seen in Figure 1, imparts strong acidic properties to the compound due to the sulfate group. Other important structural moieties include the glucose and cyano groups. Glucosinolates vary due to differences in the R group. The side group can be alkyl, branched alkyl, indole, aromatic, or unsaturated. The differences in the side chain impart different flavors to food. All glucosinolates are formed in plants from L-amino acids and sugars through a common biosynthetic pathway.

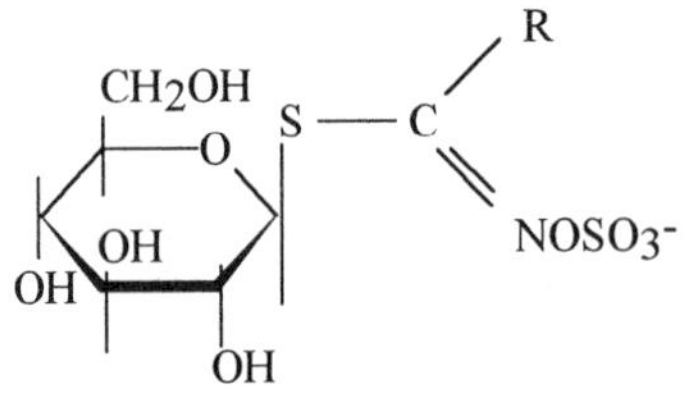

Figure 1- The structure of glucosinolates.

Glucosinolates breakdown due to the action of enzymes and heat to give many flavor compounds. Myrosinases (thioglucoside glucohydrolases) are found compartmentalized within the plant, and are released when the cell walls break through bruising, sample grinding, or during normal growth as a result of damage and death of cells. Glucosinolates are easily degraded by the myrosinase enzyme at both acidic and basic pH to give such products as isothiocyanates, nitriles, thiocyanates, oxazolidine-2-thiones, hydroxynitriles, and epithionitriles, all of which impart flavor to mustards (1).

Kale, kohlrabi, broccoli, cabbage, mustard, cauliflower, rapeseed, turnip, and rutabaga, of the *Brassica* group of cruciferous vegetables, are best known for their pungent flavor. In ancient times, these crops were thought to have medicinal power over headaches, gout, diarrhea, deafness, and stomach disorders (1). Glucosinolate breakdown products, specifically 3-indole carbinol, 3-indole acetonitrile, and 3,3'-diindolylmethane, produced by myrosinase degradation of glucobrassicin, are considered Phase II anticarcinogens (2). These compounds have also been proven to increase enzymatic activity of glutathione-S-transferase (GSH), ethoxycoumarin O-deethylase (ECD), aryl hydrocarbon hydroxylase (AHH), and epoxide hydratase (EH), all of which in turn increase the polarity of lipophilic carcinogens so that they are more readily excreted in the urine. Other benefits from the glucosinolates in *Brassica* plants include decreased growth of microorganisms, fungi, and the ability to repel most insects (3).

To date, most research into glucosinolate identification has been done by isolating glucosinolate breakdown products and then determining the original precursor (4). This method of back tracking was favored because it was inexpensive and fast. Analysis of glucosinolates has also relied heavily on GC/MS techniques on either glucosinolate degradation products such as the isothiocyanates or on the desulfated and trimethylsilyated (TMS) derivitized glucosinolates (5). Most often, in the case of MS, the isothiocyanates and other volatile components are identified rather than the glucosinolate itself (6).

In this study, negative ion electrospray mass spectrometry is used because it detects molecular anions and allows for direct identification of glucosinolates, most of which are unavailable as pure standards. We apply a novel method of reversed phase HPLC utilizing volatile ion-pairing reagents followed by negative ion electrospray mass spectrometry to semi-quantify and identify the glucosinolates in mustard seeds. Formic acid and triethylamine (TEA) were used to produce the ion-pairing agent, triethylammonium formate which interacted with the glucosinolates to increase retention and prevent their elution in the void volume.

Materials & Method

Analytically pure sinigrin (98%) was purchased from Aldrich (St. Louis, MO), as were the ion-pairing reagents, formic acid and triethylamine. Gluconasturtiin (phenethylglucosinolate) was obtained from LKT Laboratories (St. Paul., Minn). Sinapic acid (98%) was supplied by Aldrich (St. Louis, MO). HPLC-grade methanol was purchased from Fisher Scientific (Pittsburgh, PA). HPLC-grade water was generated in the laboratory through the use of a Milli-Q UV Plus Ultra-pure water system by Millipore (Milford, MA).

The external standard, phenethylglucosinolate, was carefully weighed (10.27 mg) and dissolved in 10 ml HPLC-grade water to obtain a solution with a concentration of 1.027 mg/ml. Sinigrin monohydrate (10.27 mg) was also dissolved in HPLC-grade water to obtain the same concentration as the phenethylglucosinolate. Sinapine bisulfate (98%) was purchased through Spectrum Chemical (Gardenia, CA) and was used as an external standard to quantify sinapine found in all three mustard types. All standards were filtered using a 0.45 micron Acrodisc purchased from Gelman Scientific (Philadelphia, PA). Standards were stored in amber autosampler vials under refrigeration until needed.

Brown, oriental, and yellow mustard seeds were obtained from various suppliers, Continental Grain (Alberta, Canada), North Dakota Mustard and Spice, and McCormick & Company (Hunt Valley, MD). Samples were prepared by grinding ~25 grams of mustard seed in a 250 ml polypropylene centrifuge tube, Thomas Scientific (Swedesboro, NJ), with 100 ml methanol. A Beckman Polytron was used for three minutes to insure a homogeneous grind. The samples were then centrifuged at 2700 rpm for 5 minutes. The supernatant was poured off into a 250 ml round bottom flask and then evaporated using a Rotovap set at 55°C and an rpm of 55. Samples were evaporated to a volume of approximately 2 ml methanol. The samples were then dissolved in 100 ml HPLC-grade water. All samples were filtered using a 0.45 micron nylon Acrodisc as used during standard preparation.

Analyses of mustard samples were performed by injecting a 20 ul aliquot of the sample into a Waters (Milford, MA) high-performance liquid chromatograph linked to a Waters 490 ultra-violet detector set at 235 nm. A Waters 600E System Controller and a 712 WISP Autoinjector were also used. The entire HPLC system was controlled using the Waters Millenium data acquisition software program version 2.1. The column used was a Phenomenex (Torrence, CA) 5 μm ODS(20) column measuring 250 mm X 4.6 mm with a 55 mm X 4.6 mm Waters C-18 guard column attached.

The two mobile phase solvents, methanol and water, were prepared with 0.15% triethylamine and 0.18% formic acid added as ion-pairing reagents. Both solutions were filtered using a 0.45 μm filter, sonicated for 30 minutes, and degassed with helium throughout the chromatographic procedure at a sparge rate of 50 ml/minute. The initial mobile phase was 100% HPLC-grade water. At ten minutes, the mobile phase was switched to a linear gradient of 100% water to 100%

methanol over 60 minutes. After each run the initial mobile phase conditions were set and the system allowed to equilibrate. The flow rate was kept constant at 1 ml/minute. The column temperature was held at 40°C.

Confirmation of peak identity was accomplished through the use of a Fisons (Danvers, MA) VG Platform II mass spectrometer connected to a VG Mass Lynx data system. The system was operated in the negative ion electrospray mode. The ion source temperature was set at 150°C while the cone voltage was -20V. The HPLC conditions were identical to those obtained off-line except that a Varian 9012 HPLC (Fernando, CA) was used along with a Varian 9050 UV-Vis detector.

Quantification of the nonvolatile flavor precursors, sinigrin, progoitrin, sinalbin, and sinapine was determined using external standards. A solution containing 50% phenethylglucosinolate and 50% sinigrin monohydrate was analyzed to determine the response factors of both aromatic and alkenyl glucosinolates, respectively. The ratio of response factor was 1.19, indicating that the aromatic glucosinolate was more sensitive. The concentration of the aromatic compounds, sinalbin and sinapine, were calculated using this response factor since pure standards were commercially unavailable.

Results and Discussion

Glucosinolate Quantification. Yellow, brown, and oriental mustards were analyzed for their glucosinolate content. The major compounds identified by mass spectrometry can be seen in Table I. A chromatogram of yellow mustard can be seen in Figure 2, whereas Figure 3 shows a chromatogram of brown mustard. Many of the minor peaks seen on the HPLC chromatograms were not visible by the mass spectrometry method utilized. This indicates that these compounds are uncharged and therefore are not glucosinolates. Results for the quantification of compounds in yellow mustard (*Brassica hirta*) identified by mass spec. can be seen in Table II.

Table I. Nonvolatiles Found in Yellow, Brown, and Oriental Mustards

RT	Peak #	m/z (observed)	Identification
10.5 minutes	1	358	Allylglucosinolate (Sinigrin)
11	2	388	2-hydroxy-3-butenylglucosinolate(Progoitrin)
21	3	424	p-hydroxybenzylglucosinolate (Sinalbin)
33	4	354	3,5-dimethoxy-4-hydroxycinnamoyl choline (Sinapine)
35	5	586	Sinalbin + glucose

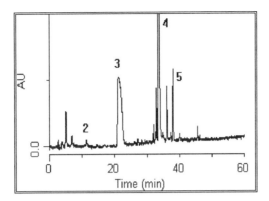

Figure 2- HPLC chromatogram of methanol extract from yellow mustard seeds
(See Table I for identifications)

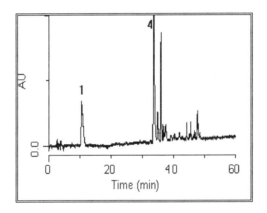

Figure 3- HPLC chromatogram of methanol extract of brown mustard seeds
(See Table I for identifications)

Table II. Glucosinolate Content of Yellow Mustard Seeds

ID #	Harvest Area	Progoitrin	Sinalbin	Sinapine
			Amount (mg/g)	
4-6	Calgary, Canada	0.51± 0.14	27.46± 1.20	27.68± 1.41
10-12	Manatoba, Canada	0.60± 0.17	27.87± 1.09	32.09+ 0.69
16-18	Alberta, Canada	0.56± 0.10	24.81± 1.92	30.60± 2.21
19-21	Saskatchewan,Canada	0.45± 0.04	23.00± 2.85	28.05± 4.15

Progoitrin, 2-hydroxy-3-butenylglucosinolate, (Figure 4) is the precursor to goitrin which has been shown to cause hypothyroidism in animals fed large amounts of cruciferous vegetables. The average level of progoitrin found in yellow mustard was 0.53 mg/g $\pm$ 0.06. The identification of progoitrin was determined directly from the negative ion electrospray mass spectrum. The mass spectrum can be seen in Figure 5. The molecular anion can be seen at m/z 388. The molecular weight of progoitrin is actually 427 (addition of potassium cation). The mass spectrometer sees only the anion when in the negative ion mode.

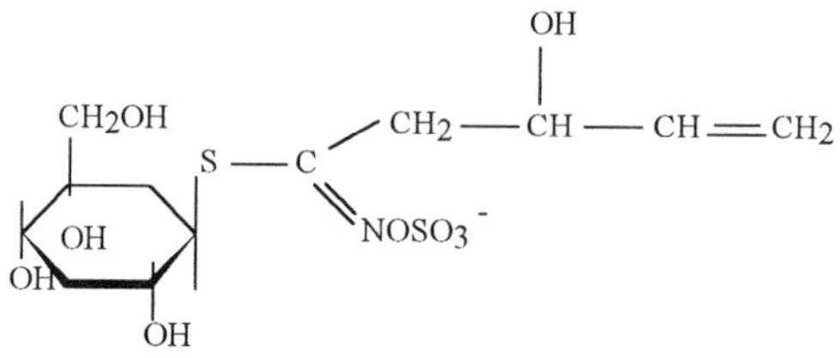

Figure 4- Structure of progoitrin anion.

Sinalbin, para-hydroxybenzylglucosinolate (Figure 6), is found at high levels in yellow mustard. The concentration of sinalbin was 25.79 mg/g $\pm$ 2.30, which is consistent with Fenwick's Critical Review of Glucosinolates and their Breakdown Products (1982) and The USDA Phytochemical data base available on the Internet (HTTP://Probe.Nalusda.Gov:8300). Sinalbin is not found in nature as a potassium salt, but rather has sinapine as the positive ion. The structure and information for sinapine is mentioned below because sinapine is also present as a peak on the chromatograms of brown, and oriental mustards.

Sinalbin was identified by mass spectrometry. The molecular anion is seen at m/z 424 (Figure 7).

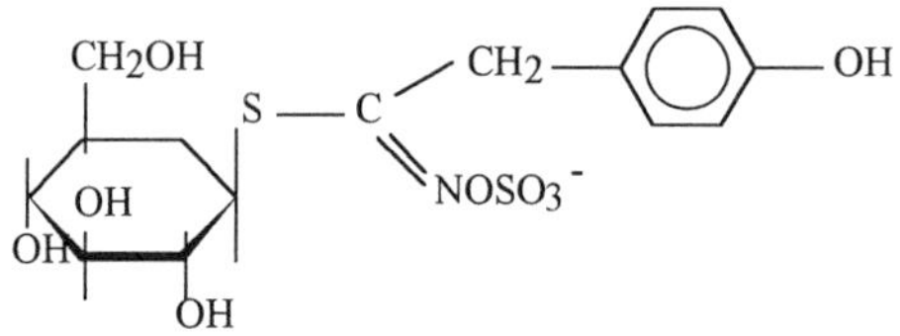

Figure 6- Structure of sinalbin anion.

The compound with a retention time of 35 minutes has been hypothesized as sinalbin plus glucose. The m/z of 586 shows 424 mass units from sinalbin and 162 from glucose. Researchers have reported the presence of glucosinolates with a second sugar in mustard and other vegetables from the *Brassica* species (4). Further research, including fractionation and NMR, must be done on this compound in order to determine the exact structure.

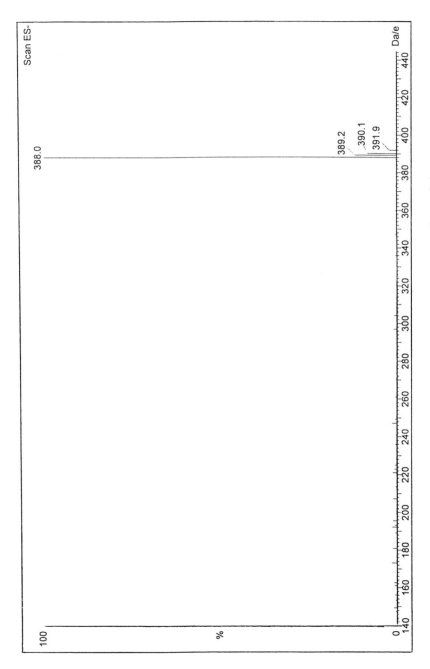

Figure 5- Negative ion electrospray mass spectrum of progoitrin.

Figure 7- Negative ion electrospray mass spectrum of sinalbin.

Brown and oriental mustard are from the same species, *Brassica juncea*, and therefore had very similar glucosinolate content. The individual results for each mustard sample can be seen in Table III.

Table III. Glucosinolate Content of Brown Mustard Seeds

ID #	Type	Harvest Area	Sinigrin	Sinapine
			Amount (mg/g)	
1--3	Oriental	Calgary, Canada	9.59± 0.94	22.05± 1.50
7--9	Oriental	Saskatawan, Canada	9.10+ 0.19	24.50± 1.09
22--24	Oriental	Saskatchewan,Canada	9.31± 0.52	25.00± 1.97
25--27	Oriental	Alberta,Canada	8.16± 0.36	20.24± 1.77
13--15	Brown	Saskatchewan,Canada	8.29± 0.26	24.20± 0.92
28--30	Brown	Alberta, Canada	8.80± 1.97	14.36± 2.70
31--33	Brown	Saskatchewan,Canada	8.20± 0.95	12.98± 1.78

The major compound in brown and oriental mustards is sinigrin, commonly known as allylglucosinolate, (Figure 8) with a retention time of approximately 10.5 minutes. Sinigrin degrades to allylisothiocyanate (AITC) by the myrosinase enzyme in the presence of water to give mustard its pungency. The concentration of sinigrin was found to be 9.04 mg/g $\pm$ 0.62 in oriental mustard and slightly lower than that for brown. The CV of the sinigrin results for all twenty one samples in the *Brassica juncea* group was 10.69%. The identity of sinigrin was proven by negative ion electrospray mass spectrometry when the MS showed a peak with a mass to charge ratio of 358 as seen in Figure 9. The negative ion is seen by the mass spectrometry method used, whereas the positive ion, potassium, is not observed. The actual molecular weight of sinigrin is 397.

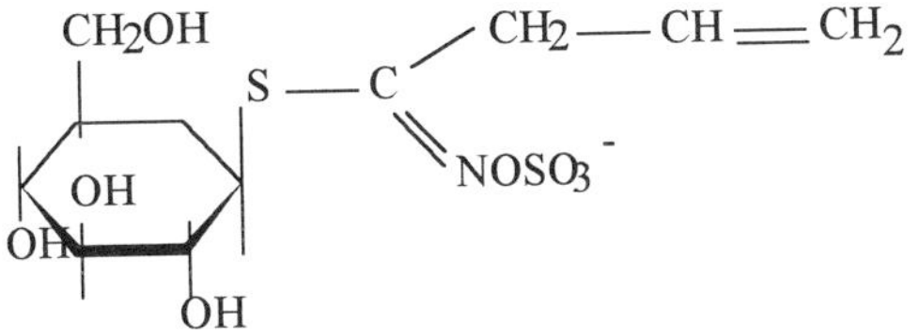

Figure 8- Structure of sinigrin anion.

Sinapine, a choline ester of sinapic acid (Figure 10) was identified in yellow, brown, and oriental mustard. This compound, 3,5-dimethoxy-4-hydroxy cinnamoyl choline, is common in the *Brassica* group. Researchers have shown that although sinapine is the positive ion to sinalbin, the glucosinolate and phenolic choline ester contents are not correlated with one another (7). For this reason sinapine is seen in both brown

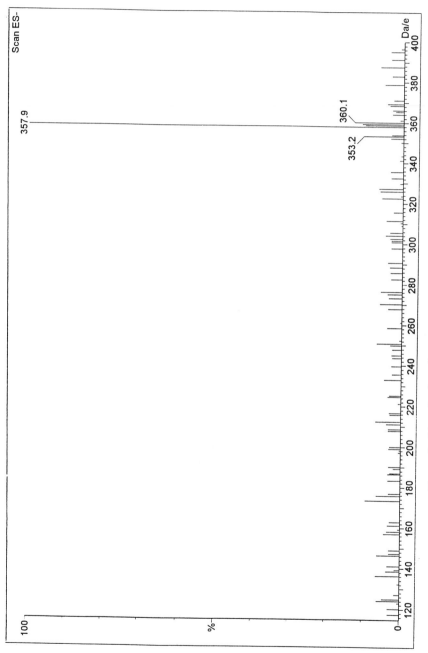

Figure 9- Negative ion electrospray mass spectrum of sinigrin.

and oriental mustards although sinalbin is not present.   The negative ion electrospray mass spectrum of sinapine using a low cone voltage of 20 V, showed an ion at m/z 354. The peaks seen above the m/z 354 peak in Figure 11 are 46 mass units higher, which is the compound forming adducts with formic acid, the mobile phase modifier.   Since very little fragmentation occurs at such a low cone voltage, a second mass spectrum was run at a higher cone voltage, 120 volts, to give fragmentation. This data and a NMR of the fractionated peak showed that the compound was an adduct of sinapine and formate, 310 mass units coming from the sinapine cation and 45 amu from the formate anion. The [M-H]⁻ of this complex occurs at m/z 354. Interestingly, because of the ion-pairing, the sinapine did not give a spectrum in the positive ion mode.

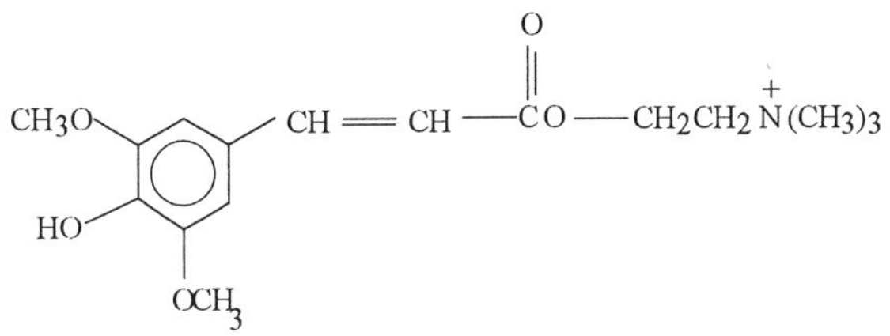

Figure 10- Structure of sinapine.

Conclusion

To date, the major nonvolatiles in mustard seeds have been isolated, identified, and quantified by reversed phase HPLC using ion-pairing reagents and negative ion electrospray mass spectrometry.   Accuracy and precision of the method was determined by both spiking studies and replicate analysis of individual samples, respectively.   The accuracy was determined by spiking both brown and yellow mustard samples with sinigrin and calculating the percent recovery. Recovery was often greater than 100%. Precision was determined by running all samples in triplicate and calculating the standard deviation and coefficient of variation. CV of most samples was less than 10%. This data indicates that this method works well for identifying intact glucosinolates and therefore eliminates the need for desulfating and derivitizing glucosinolates as seen in previous studies (5). Use of mass spectrometry allows identification without glucosinolate standards, most of which cannot be purchased.

Acknowledgments

The author gratefully acknowledges the work of Dr. Elaine M. Fukuda, CAFT, Rutgers University and Dr. Janet M. Deihl, Nabisco, Inc. This is New Jersey Agricultural Experiment Station Publication # D10570-96-2.

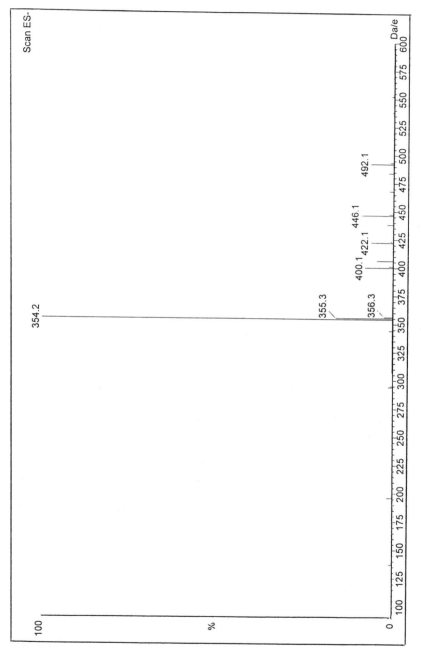

Figure 11- Negative ion electrospray mass spectrum of sinapine.

Literature Cited

1. Fenwick, G. R.; Heaney, R. K.; Mullin, W. J., *CRC Crit. Rev. Food Nut.*, **1982**, 18, 123-201.
2. Bradfield, C. A.; Bjeldanes, L. F. *Fd. Chem. Toxic.* **1984**, 22, 977-982.
3. Delaquis, P.J.; Mazza G. *Food Tech,* **1995,** 11, 73-84.
4. Oleson, O.; Sorensen, H., *J. Amer. Oil Chem. Soc.* **1981**, 857-865.
5. Helboe, P.; Olsen, O.; Sorensen, H., *J. Chrom.,* **1980**, 197, 199-205.
6. Kjaer, A.; Ohashi, M.; Wilson, J. M., Djerassi, C. *Acta Chemica Scandinavica,* **1963**, 17, 2143-2154
7. Bouchereau, A.; Hamelin, J; Lamour, I.; Renard, M.; Larher, F.; *Phytochemistry,* **1991**, 30 (6), 1973-1881.

Chapter 12

Reasons for the Variation in Composition of Some Commercial Essential Oils

Chi-Kuen Shu and Brian M. Lawrence

R. J. Reynolds Tobacco Company, 950 Reynolds Boulevard, Winston-Salem, NC 27102

The reasons which cause the compositional changes for some commercial essential oils are due to (1) the effect of extrinsic conditions, (2) the effect of interspecific and infraspecific differences, (3) the effect of ontogeny, (4) the effect of processing parameters, and (5) the effect of adulteration. Using examples, each effect is discussed.

Over the years it has long been known that commercially available essential oils vary from season to season and from geographic source to geographic source. However, this question of variation has not been carefully examined across a range of oils to determine why such variations take place. Usually it is explained that changes from one season to another result in compositional changes; however, this is a little too simple to explain some of the compositional changes that are encountered.

As a result, we would like to discuss some basic reasons why compositional changes are encountered using examples of the magnitude of such changes. The effects that will be discussed are:

1. The effect of extrinsic conditions (climate/geographical origin).
2. The effect of interspecific and infraspecific differences (between and within species).
3. The effect of ontogeny (growth stage of plant material harvested).
4. The effect of processing parameters (dry vs. wet plant material on distillation changes).
5. The effect of adulteration (the addition of synthetic or other natural components to "stretch" an oil.

The Effect of Climate and Geographic Origin on Oil Composition.

In 1989 Lawrence et al. (1) showed that peppermint oil, which is produced from clonally reproduced plants that are grown in different regions of the United States, could be readily differentiated based on their area of cultivation and oil production. The areas of commercial production of these oils are the Midwest, Idaho, Yakima-Kennewick Valley in Washington, and the Madras and Willamette valleys of Oregon. It was proved that geographical location and microenvironment were influential factors which affect, albeit slightly, the composition of the oils produced in each region. Examination of the quantitative data of the most constituents revealed that it was extremely difficult to differentiate the oils because of the similarity in the data. As a result, selected component ratios were used to prove differences between the oils. To simplify the component ratio data obtained from ca. 50 samples of oil obtained from each growing region, they were presented as polygonal representations. Such a presentation of data can be thought of as being pictorial pattern recognition of the oils produced in each region.

The components ratios selected to differentiate between the oils can be seen in Table I. Many of these ratios have multipliers that were used to ensure that the polygons formed would fit the scale of the circular graph. What this means is that only oils produced in a specific region will have their component ratio data fit within the polygon. Oils whose component ratio data does not fit within the polygon cannot be a pure oil produced in that specific region. The pattern recognition of the five different types of peppermint oil can be seen in Figures 1-5.

To further use this concept to differentiate between other oils produced in the United States from different regions we attempted to differentiate between Native spearmint oils (ex. <u>Mentha spicata</u> L.) grown in the Midwest and in the Farwest. Similarly, we also attempted to differentiate between Scotch spearmint oils (ex. <u>Mentha gracilis sole</u>), which are also produced in the Midwest and the Farwest. We could not use tighter regional differences between the oils because the oils are not sold according to the state or valley in which they are produced, only by the broad regions of Midwest and Farwest.

For both spearmint oils the component ratios used can be seen in Table II. Again, multipliers were used to ensure that the component ratio data for both spearmint oils would fit on the same circular graph. The results of the component ratio data that were obtained for 20 samples of each oil type plus 10 samples of a second cutting oil obtained from Farwest Native spearmint can be seen in Figures 6-10. The compounds chosen for component ratios were those whose raw analytical data varied so that a differentiation between Scotch and Native spearmint oils and 2nd cutting Native spearmint oils could be made. For example, Scotch spearmint oil is slightly richer in limonene, 3-octanol, menthone and carvone as compared to Native spearmint oil, which in turn, is slightly richer in mycrene, 1-8-cineole, 3-octyl acetate, trans-sabinene hydrate, β-bourbonene and trans-carveol than Scotch spearmint oil. However, because it is difficult to look at raw analytical data and readily see slight differences, the use of

Table I. Component Ratios Selected for Peppermint Oil Differentiation According to Origin

1.	1,8-Cineole/Limonene
2.	1,8-Cineole/Menthofuran x 1/2
3.	1,8-Cineole/Menthyl Acetate
4.	1,8-Cineole/Menthol x 50
5.	1,8 Cineole/Menthone x 25
6.	Menthofuran/Menthone x 100/6
7.	Menthofuran/Menthol x 50
8.	Menthofuran/Menthyl Acetate x 5
9.	Menthofuran/Limonene
10.	Menthone/Menthyl Acetate x 1/2
11.	Menthol/Menthone
12.	Menthol/Menthyl Acetate x 1/2

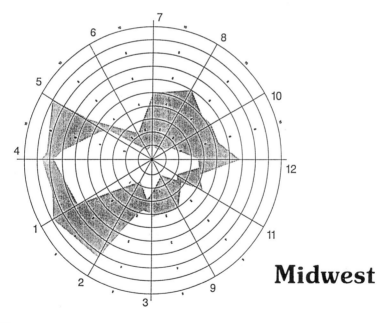

Figures 1-5   Pattern Recognitions of the Five Types of Peppermint Oil

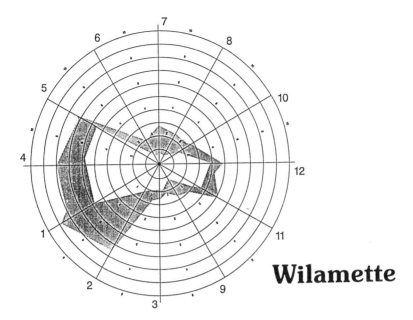

Figure 2.

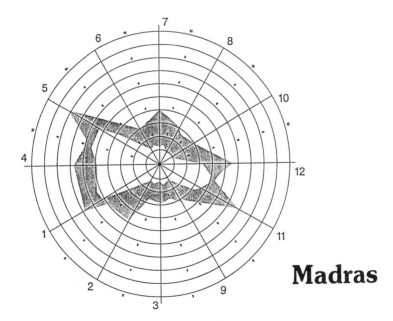

Figure 3.

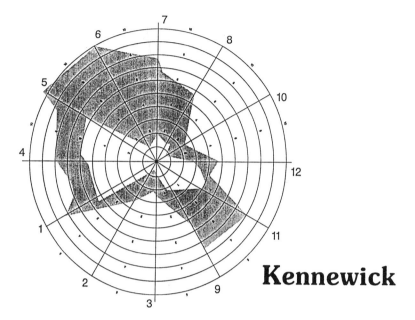

Kennewick

Figure 4.

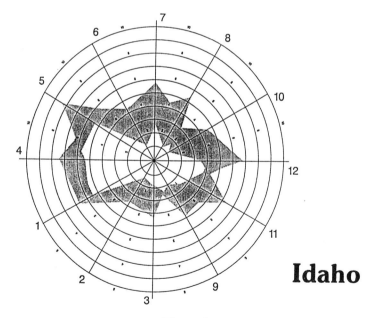

Idaho

Figure 5.

Table II. Component Ratios Used to Differentiate between Midwest and Farwest Native Spearmint Oils and Scotch Spearmint Oils

1.	Limonene/Myrcene x 1/2
2.	Limonene/1,8-Cineole x 1/2
3.	Limonene/3-Octanol x 1/5
4.	Limonene/β-Bourbonene x 1/5
5.	Myrcene/β-Bourbonene x 2
6.	1,8-Cineole/Myrcene x 5
7.	Carvone/Myrcene x 1/10
8.	Carvone/β-Bourbonene x 1/10
9.	3-Octanol/3-Octyl acetate x 1/2
10.	1,8-Cineole/trans-Sabinene hydrate x 1/2
11.	cis-Carveol/trans-Carveol x 1/10
12.	Carvone/trans-Carveol x 1/150

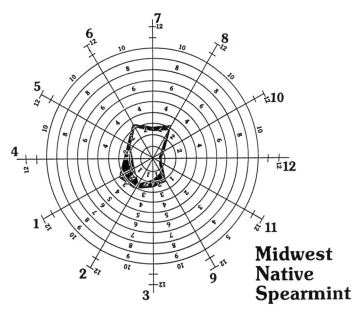

Midwest Native Spearmint

Figures 6-10   Pattern Recognitions of the Five Types of Spearmint Oil

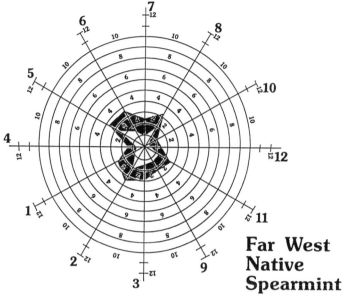

Far West Native Spearmint

Figure 7.

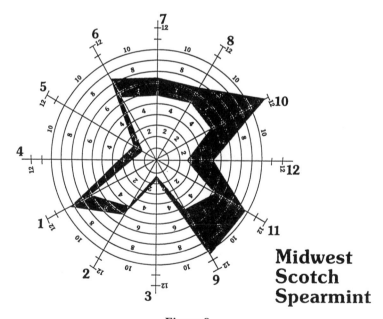

Midwest Scotch Spearmint

Figure 8.

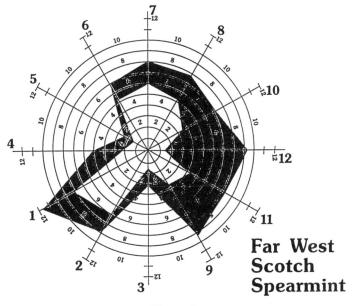

Far West
Scotch
Spearmint

Figure 9.

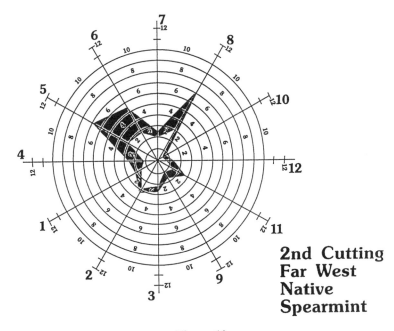

2nd Cutting
Far West
Native
Spearmint

Figure 10.

component ratios magnifies these differences so that they can both be readily discernible and can be easily displayed pictorially.examination of the limonene/myrcene ratios for each different spearmint oil, we have 1.81-3.39 (Midwest Native), 0.88-2.68 (Farwest Native), 1.60-2.02 (2nd cutting Farwest Native), and 6.99-9.27 (Midwest Scotch), 8.12-11.48 (Farwest Scotch). From these data it can be seen that it is easy to differentiate Native spearmint oil from Scotch spearmint oil. Using the eleven other component ratios merely makes this differentiation task easier. When this data is presented as a polygonal representation, this differentiation is readily evident from these pattern recognition figures (Figure 6-10) that:

(a)   Native and Scotch spearmint oils can be readily differentiated.

(b)   Farwest and Midwest Scotch and Native spearmint oil can be readily differentiated.

(c)   A second cutting of oil of Farwest Native spearmint can be readily differentiated from all of the other spearmint oils.

The Effect of Interspecific and Infraspecific Differences on Oil Composition.

In the previous above example, it was shown that two spearmint oils, which were produced from clonally reproduced distinctly different species of Mentha, produced oils whose compositions were similar but could be differentiated; however, the detailed oil compositional data were not presented. To address this point, the analysis of oils that are known commercially as cedarwood will be the next subject of discussion.

Around the world a number of oils are produced that are described as being cedarwood oil. Each of these oils is produced from a different species as can be seen in Table III. The oils produced from Atlas cedar, Himalayan cedar, Incense cedar, Lebanon cedar, East African cedar, Oregon cedar, Mulanje cedar, Alaskan cedar, Eastern white cedar and Western red cedar are known to possess different chemical compositions. The major components of these oils can be seen in Table IV. In contrast, the Texas, Virginia, Chinese , and Atlantic cedarwood oils, although produced from different species, all possess compositions that are quite similar (Table V) Using the hypothesis that it is easier to look at component ratios rather than raw data, the components that show most variation between the four cedarwood oils were α-cedrene, thujopsene, β-himachalene and cedrol. As a result, the component ratios were examined rather than the raw data (Table VI). A comparison between these data reveals that Virginia cedarwood oil can be clearly differentiated by the thujopsene/α-cedrene and β-himachalene/cuparene ratios. It is more difficult to differentiate Texas and Chinese cedarwood; however, from the thujopsene/cedrol ratios it is possible. Finally Atlantic cedarwood oil possesses a unique β-himachalene/cuparene ratio, thereby differentiating it from the other oils.

To further discuss this question of interspecific differences in composition, lemongrass oil will be used as a second example. There are two forms of lemongrass oil produced commercially, one produced in Asia from Cymbopogon flexuosus (Nees ex Steud.) Wats. and the other produced in the

Table III. Various "Cedars" that have been used to produce cedarwood oil

1.	Virginia Cedar	Juniperus virginiana L.
2.	Texas or Mexican Cedar	Juniperus ashei Buchholz
3.	Western Red Cedar	Thuja plicata Donn ex. D. Don
4.	Northern or Eastern White Cedar	Thuja occidentalis L.
5.	Kenyan or East African Cedar	Juniperus procera Hochst ex. Endl.
6.	Chinese Cedar	Chamaecyparis funebris (Endl.) Franco (syn. Cupressus funebris Endl.)
7.	Alaskan Cedar	Chamaecyparis nootkatensis (D. Don) Spach
8.	Oregon Cedar	Chamaecyparis lawsoniana (Andr.
9.		Murray)
10.	Mulanje Cedar	Widdringtonia cupressoides (L.) Endl.
11.	Atlas Cedar	Cedrus atlantica G. Manett
12.	Lebanon Cedar	Cedrus libani A. Rich.
13.	Himalayan Cedar	Cedrus deodara (Roxb.) Loud.
14.	Atlantic Cedar	Chamaecyparis thyoides (L.) B.S.P.
15.	Incense Cedar	Libocedrus decurrens Torrey

Table IV. Major Components in Various Cedarwood Oils

3.	methyl thujate T-muurolol tropolones	10.	α-himachalene β-himachalene γ-himachalene
4.	occidentatol occidol	11.	α-torosol β-torosol (E)-atlanton-6-ol
5.	cedrol	12.	(Z)-α-atlantone (E)-α-atlantone
7.	γ-cadinene 1 (10)-cadinen-4β-ol	14.	p-methoxythymol carvacrol tropolones
8.	δ-cadinene α-pinene borneol		
9.	thujopsene cedrol β-himachalene		

Table V. Main Components of Four Cedarwood Oils

Compound	V	T	C	A
α-cedrene	38.0	22.6	33.8	20.5
β-cedrene	9.2	5.5	8.1	5.0
thujopsene	23.4	46.8	36.0	31.7
α-himachalene	-	0.8	0.7	-
β-chamigrene	1.4	-	-	1.5
β-selinene	-	-	-	0.8
β-himachalene	2.1	1.1	0.7	0.2
α-chamigrene	-	1.5	1.6	-
ar-curcumene	-	trace	0.3	-
cuparene	0.9	1.9	2.2	3.5
cedrol	12.3	12.2	8.5	10.4
widdrol	1.9	1.1	1.1	1.4
α-bisabolol	trace	trace	trace	0.3

Cedarwood Oils: V = Virginia   C = Chinese
　　　　　　　　　　T = Texas   A = Atlantic

Table VI. Component Ratios Used to Differentiate Between Thujopsene-rich
　　　　　　　Cedarwood Oils

Compound Ratios	V	T	C	A
thujopsene/α-cedrene	2 - 3	8 - 9	4 - 5	6.5 - 7.5
thujopsene/cedrol	1.7 - 2.1	3.6 - 4.0	4.0 - 4.4	2.9 - 3.1
β-himachalene/cuparene	2.2 - 2.4	0.5 - 0.7	0.2 - 0.4	0.01 - 0.1

Cedarwood Oils: V = Virginia   C = Chinese
T = Texas   A = Atlantic

Table VII. Chemical Composition of E. Indian Lemongrass Oil

Compound	Indian I	Indian II	Bhutanese
α-thujene	trace	-	2.8
α-pinene	-	0.4	5.4
camphene	0.5	1.0	10.1
myrcene	trace	-	10.3
limonene	2.7	3.8	8.3
4-nonanone	0.5	0.3	trace
6-methyl-5-hepten-2-one	2.6	2.5	3.2
citronellal	0.8	0.9	0.8
linalool	1.4	1.1	2.1
β-caryophyllene	1.3	0.9	0.4
neral	32.6	34.1	18.5
geranial	40.8	44.3	22.7
geranyl acetate	3.9	2.2	3.4
citronellol	0.5	0.6	0.6
geraniol	4.9	3.8	2.2

New World from Cymbopogon citratus (DC) Stapf. Authentic oils obtained from each species were subjected to analysis and by GC/MS. A summary of the differences between their compositions can be seen in Table VII and VIII. Although the differences between these oils appear to be major because of the high limonene content of the West Indian oils, it is easier and more prudent to use component ratios to differentiate between the oil as shown in Table IX. It is believed that examination of the hydrocarbons and oxygenated compounds separately is a good way to examine the authenticity of an oil.

The existence of morphologically identical plants which possess oils that have differing chemical compositions is not a new concept. In fact, the existence of infraspecific differences is widespread in the Labiatae and Compositae families, but it is not limited to them. A few years ago Lawrence (2) showed that Ocimum basilicum L. can contain oils that possess a variety of compositions. It was also found that oils contained constituents that were biosynthesized either via the shikimic acid pathway, or the mevalonic acid pathway, or both (Figure 11). A summary of the data obtained during the analysis of more than 200 separate O. basilicum plants revealed that they possessed oils that, not only contained components from single pathways or dual pathways, but also within these groupings a wide quantitative variation of constituents was also observed. A summary of these data can be seen in Tables X and XI.

The Effect of Growth Stage of the Harvested Plant on the Composition of Its Oil.

Although the effect of growth stage of the plant on oil composition should be well known, oils produced commercially are not always produced year after year from plants that were harvested at their same growth stage. Two examples will be used to illustrate this point.

First, oils produced from Coriandrum sativum L. differ chemically quite drastically (3) as can be seen from the data presented in Table XII. The oils that are known as cilantro oil are generally those produced from plants at stages 1 and 2; however, the oil yield is higher the later the plant is harvested. Nevertheless, using the ratio between decanal & (E)-2-decenal and linalool and (E)-dodecenal, a reproducible cilantro oil can be obtained (Table XIII).

A second example of a change in chemical composition as the plant matures can be found with the oil of Tagetes minuta L. (4). Examination of the main acyclic hydrocarbon and the acyclic ketones (Table XIV and Figure 12) will allow the user to select an oil that is most liked. From these data, component ratios can be used as a quality check for oil reproducibility. The reason why oils can fall outside the desired component ratio can be understood based on the data presented in Table XIV.

The Effect of Processing Parameters on Oil Composition

In 1985, Schmaus and Kubeczka (5) examined the influence of oil isolation conditions on the composition of linalyl acetate-rich oils. They found that the increased water content in plants distilled when fresh, resulted in a corresponding

Table VIII. Chemical Composition of W. Indian Lemongrass Oil

Compound	Guatemala I	Guatemala II
limonene	12.3	15.9
4-nonene	1.8	2.4
6-methyl-5-hepten-2-one	3.0	4.4
citronellal	0.9	0.6
linalool	1.8	2.1
β-caryophyllene	2.2	1.3
neral	30.5	28.6
geranial	37.3	37.1
geranyl acetate	1.7	1.1
citronellol	0.4	0.2
geraniol	4.6	2.6

Table IX. Component Ratios Used to Differentiate between E. Indian and W. Indian Lemongrass Oil

Component/Ratio	E. Indian	W. Indian
methyl heptenone/4-nonanone	5.0 - 9.5	1.0 - 2.0
geraniol/geranyl acetate	1.2 - 1.8	2.2 - 3.0
citronellal/citronellol	1.1 - 1.7	2.0 - 3.2
neral/limonene	8.5 - 12.5	1.5 - 3.0

Table X. Major Components of Various *Ocimum basilicum* Selections (Single Biosynthetic Pathways)

Compound	1(a)	1(b)	2(a)	2(b)	3(a)	4(a)
1,8-cineole	3.2	4.6	1.4	2.7	2.4	3.4
(Z)-β-ocimene	0.1	0.3	0.2	0.1	trace	trace
(E)-β-ocimene	0.6	5.4	1.0	2.2	0.4	trace
camphor	0.2	1.8	0.3	4.3	0.9	1.8
linalool	0.5	4.8	83.9	51.6	0.6	6.2
methyl chavicol	85.3	49.0	0.4	0.7	1.5	5.0
geraniol	trace	trace	0.3	22.8	trace	trace
methyl eugenol	1.9	19.2	trace	trace	67.7	0.3
(E)-methyl cinnamate	trace	trace	trace	trace	0.3	64.0
eugenol	trace	0.2	0.5	trace	2.0	trace

Legend:Chemotypes of *Ocimum basilicum* with a Single Biosynthetic Pathway:

Type 1.	(a)	methyl chavicol-rich
	(b)	methyl chavicol>methyl eugenol
Type 2.	(a)	linalool-rich
	(b)	linalool>geraniol
Type 3.	(a)	methyl eugenol-rich
Type 4.	(a)	methyl cinnamate-rich
Type 5.	(a)	eugenol-rich (not found)

Table XI. Major Components of Various *Ocimum basilicum* Selections (Dual Biosynthetic Pathways)

Compound	1(c)	1(c)'	2(c)	2(c)'	2(d)	2(e)	3(b)
1,8-cineole	5.7	3.8	4.8	1.9	3.8	2.8	7.7
(z)-β-ocimene	0.1	1.2	0.4	0.6	1.4	0.2	0.1
(E)-β-ocimene	2.0	2.4	1.3	1.0	2.5	1.0	3.1
camphor	1.5	1.3	0.5	0.2	4.9	1.5	0.8
linalool	22.9	23.0	44.9	51.0	30.2	34.2	21.2
methyl chavicol	52.1	39.3	32.4	21.0	2.1	1.2	0.7
geraniol	trace	0.2	0.3	0.2	0.4	0.2	trace
methyl eugenol	0.1	0.3	0.1	0.1	10.2	0.1	38.4
(E)-methyl cinnamate	trace	15.5	0.1	trace	trace	27.8	trace
eugenol	trace	trace	3.7	13.9	14.8	4.9	5.1

Legend: Chemotypes of *Ocimum basilicum* with a Dual Biosynthetic Pathway:

Type 1.　　(c) methyl chavicol>linalool
　　　　　　(c)'methyl chavicol>linalool>methyl cinnamate
Type 2.　　(c) linalool>methyl chavicol
　　　　　　(c)'linalool>methyl chavicol>eugenol
　　　　　　(d) linalool>eugenol>methyl eugenol
　　　　　　(e) linalool>methyl cinnamate
Type 3.　　(b) methyl eugenol>linalool

Table XII. Comparative chemical composition of *Coriandrum sativum L.* at Various Stages of Maturity

Compound	Stages of Plant Maturity[2]					
	1	2	3	4	5	6
Octanal	1.20	1.20	0.85	0.66	0.44	0.35
Nonanal	0.51	0.20	0.11	0.05	0.05	0.08
Decanal	30.0	18.09	11.91	6.30	6.24	1.61
Camphor	0.08	trace	0.52	1.26	2.18	2.44
(E)-2-Decenal	20.6	46.5	46.5	40.6	30.2	3.9
Dodecanal	3.30	1.67	0.96	0.64	0.52	0.41
(E)-2-Undecenal	2.56	21.7	1.39	--	--	--
Tridecanal	3.07	1.87	2.02	0.92	1.08	0.46
(E)-2-Dodecanal	7.63	8.14	5.95	4.59	4.78	2.49
Tetradecanal	0.68	0.30	0.12	0.15	0.11	0.15
(E)-2-Tridecenal	0.49	0.21	0.14	0.09	0.09	0.13
(E)-2-Tetradecenal	4.45	2.57	1.73	1.53	1.59	1.73
Linalool	0.34	4.27	17.47	30.05	40.88	60.37
Geraniol	0.19	0.11	0.35	0.71	0.93	1.42
Geranyl acetate	4.17	0.78	0.76	0.69	0.69	0.66

[2] Stages of maturity: 1 = floral initiation; 2 = nearly full flowering; 3 = full flowering, primary umbel young green fruit; 4 = past full flowering 50% flower, 50% fruit; 5 = full green fruit; 6 = brown fruit on lower umbels, green fruit on upper umbels; -- = not detected

Table XIII. Component Ratios of Cilantro Oil Constituents as a measurement of oil quality

Component Ratio	Stages of Maturity					
	1	2	3	4	5	6
decanal/(E)-2-decanal	1.5	0.4	0.3	0.2	0.2	0.4
linalool/(E)-2-dodecenal	0.0	0.52	2.94	6.55	8.55	24.3

Table XIV. Change in percentage composition of *Tagetes minuta* oil during life cycle of plant

Development Stage	(Z)-β-ocimene	Dihydro-tagetone	(Z)-	(E)-	(Z)-	(E)-
			Tagetone		Tagetenone	
vegetative	16.9	46.4	22.4	3.2	1.5	2.0
buds visible	17.8	51.3	18.5	1.4	4.4	3.0
buds opening	21.9	36.0	24.1	1.3	6.9	1.5
flowers opening	33.3	30.0	17.5	2.0	8.6	1.9
full flower	35.3	30.5	16.9	2.2	6.9	1.8
immature fruit	41.3	24.1	8.2	0.3	18.2	2.0
mature fruit	45.9	14.8	7.5	0.4	20.4	3.6

Table XV. Comparative Chemical Composition of French and Russian Clary Sage Oils

Compound	French Oils		Russian Oil
	A (dried)	B (fresh)	
α-pinene	0.20 - 0.21	0.08	0.02
β-pinene	0.34 - 0.38	0.98	0.22
limonene + 1,8-cineole	0.19 - 0.20	0.29	0.41
(Z)-β-ocimene	0.16	0.40	0.09
(E)-β-ocimene	0.21 - 0.23	0.72	0.06
linalool	8.88 - 8.98	28.48	14.43
α-terpineol	0.20 - 0.27	2.64	2.17
nerol	0.05 - 0.06	0.56	0.42
geraniol	0.11	1.54	0.98
linalyl acetate	72.47 - 74.18	50.86	63.18
neryl acetate	0.13 - 0.17	0.75	1.03
geranyl acetate	0.30 - 0.40	1.42	2.07
β-caryophyllene	1.85 - 1.89	1.65	2.02
germacrene D	3.29 - 4.09	3.54	0.36
caryophyllene oxide	0.35 - 0.40	0.16	0.16
sclareol	1.21 - 1.97	0.97	1.80

increase in distillation time and a decrease in pH causing partial hydrolysis of linalyl acetate followed by partial acid catalyzed degradation of linalool resulting in an increase of (E)- and (Z)-β-ocimene, limonene, terpinolene, α-terpineol, geraniol, neryl acetate, and geranyl acetate. In 1985, Doré and Jaubert (6) presented an example of these effects as it applies to clary sage oil (Table XV). Similar results were presented by Lawrence (7) (Table XVI) where the U.S. oil was produced from fresh plant material, while the other two oils were produced from dried material. The effects described by Schmaus and Kubeczka (5) are very evident in these results. A processing parameter that needs to be mentioned is storage. If an oil is stored under poor conditions such as in a half-filled drum in which some dissolved water has deposited causing some minute rusting in the drum, oxidation of the oil can and does occur. A self-explanatory example of the oxidation of coriander oil can be seen in Table XVII.

The Effect of Adulteration on the Composition of an Oil

Two examples of oil adulteration will be discussed namely, lemongrass oil and peppermint oil. Examination of the data presented in Table XVIII shows the analysis of three commercial samples of lemongrass oil. Sample Nos. 2 and 3 are grossly adulterated because they were found to contain ca. 20% of either diacetone alcohol or triacetin as an adulterant. In contrast, Sample No. 1 has to be examined from the standpoint of component ratios. A comparison between the component ratios of E. Indian, W. Indian lemongrass oil and Sample No. 1 (Table XIX) reveals that the oil is not a true example of either lemongrass oil; therefore, it must be adulterated.

Over the past few years the production of Indian peppermint oil has increased, both in the Punjab area as well as in the Bareilly region of Uttar Pradesh. As this oil is substantially lower in price than North American peppermint oil, an economic temptation exists to mix Indian peppermint oil with one of the U.S. peppermint oils and sell it as U.S. oil. As a result, it was decided to determine if this practice at the 10% adulteration level could be readily detected.

Initially, 18 oils from both regions were analyzed by GC/MS to ensure that there were no unusual constituents that would interfere with further studies. The component ratios shown in Table I were calculated, and these were plotted on a circular graph to show the typical pattern of an Indian peppermint oil (Figure 13). From the individual Indian peppermint oil samples, a composite was made, and it was added at a 10% inclusion rate into a composite oil of the Midwest, Idaho, Kennewick, Madras, and Willamette peppermint oils that were analyzed earlier. These adulterated oils were then analyzed by GC and the important peak identities were confirmed by GC/MS. The component ratios of each adulterated oil were calculated and were plotted against the original pattern recognition polygonal representations of the oils from each region oil (Figures 14-18).

As can be seen, the data plot for the U.S. oil containing 10% of the Indian composite oil fell outside of the polygons, thereby confirming that each oil was adulterated.

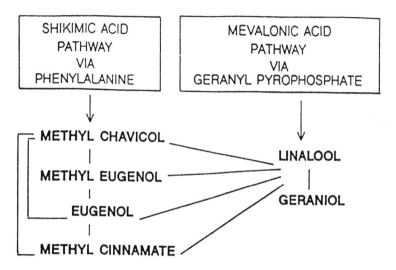

Figure 11   Dual Biosynthetic Pathway Found in *Ocimum basilicum*

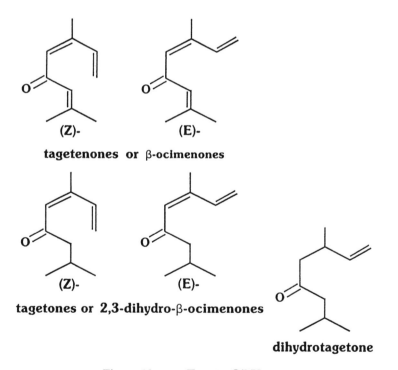

Figure 12   **Tagetes Oil Ketones**

Table XVI. Comparative Chemical Composition of Commerical Clary Sage Oil

Compound	U.S. Oil	French Oil	Russian Oil
α-pinene	0.17 - 0.22	0.1 - 0.3	0.2 - 0.3
myrcene	1.25 - 1.71	0.1 - 0.2	0.3 - 0.5
limonene	0.40 - 0.77	0.1 - 0.2	0.1 - 0.2
(Z)-β-ocimene	0.42 - 0.70	trace	trace - 0.2
(E)-β-ocimene	0.43 - 1.36	0.1 - 0.2	0.1 - 0.4
linalool	20.29 - 28.63	9.0 - 16.0	10.4 - 19.3
linalyl acetate	44.9 - 53.4	49.0 - 73.6	45.3 - 61.8
β-caryophyllene	0.86 - 1.31	1.4 - 1.6	1.1 - 1.8
α-terpineol	1.05 - 3.05	0.2 - 0.6	1.2 - 2.5
germacrene D	2.63 - 3.55	1.6 - 2.0	0.7 - 2.0
neryl acetate	1.00 - 1.67	0.2 - 0.3	0.4 - 0.6
geranyl acetate	1.93 - 3.24	0.3 - 0.5	0.8 - 1.2
nerol	0.62 - 1.15	trace - 0.1	0.3 - 0.5
geraniol	1.67 - 3.26	0.1 - 0.3	0.6 - 1.2
caryophyllene oxide	0.21 - 0.27	0.3 - 0.5	0.5
sclareol	0.17 - 0.44	0.1 - 0.2	0.1 - 0.2

Table XVII. The Effect of Storage (Oxidation) on Coriander Oil

Compound	Good Oil	Oxidized Oil
p-cymene	1.7 - 3.8	1.9
cis-linalool oxide (furan)	0.2 - 0.5	14.1
trans-linalool oxide (furan)	0.2 - 0.4	12.3
camphor	2.1 - 4.4	6.3
linalool	70.3 - 77.1	38.1
4-methyl-5-hexen-4-olide	--	3.1
cis-linalool oxide (pyran)	--	1.5
trans-linalool oxide (pyran)	--	1.4
p-cymen-8-ol	--	0.6

Table XVIII. Chemical Composition of Adulterated Lemongrass Oil

Compound	1	2	3
myrcene	1.8	trace	trace
limonene	--	1.4	1.6
4-nonanone	--	0.6	0.4
6-methyl-4-heptenone	3.4	1.9	1.6
diacetone alcohol	--	21.3	--
triacetin	--	--	25.8
citronellal	0.6	0.5	0.5
linalool	2.9	1.3	1.5
β-caryophyllene	0.5	1.0	0.9
neral	36.2	26.3	24.1
geranial	45.5	32.7	31.0
geranyl acetate	0.7	4.1	3.8
citronellol	0.3	0.3	0.3
geraniol	2.6	6.0	5.6

Table XIX.Component Ratios Used to Differentiate Between E. Indian
and W. Indian Lemongrass Oil and an Adultrated Oil (Sample No. 1)

Component Ratio	E. Indian	W. Indian	Sample No. 1
methyl heptenone/4-nonanone	5.0 - 9.5	1.0 - 2.0	--
geraniol/geranyl acetate	1.2 - 1.8	2.2 - 3.0	3.7
citronellal/citronellol	1.1 - 1.7	2.0 - 3.2	2.0
neral/limonene	8.5 - 12.5	1.5 - 3.0	--

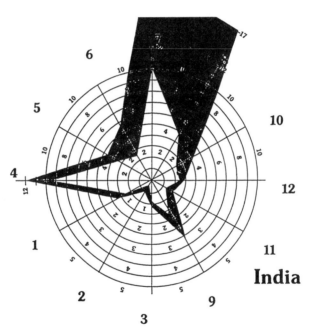

Figure 13   Pattern Recognition of an Indian Peppermint Oil

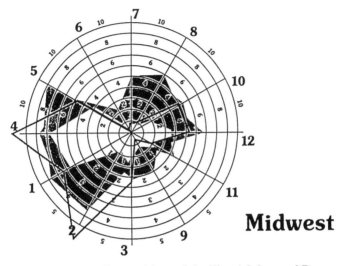

Figures 14-18 Pattern Recognitions of the Five Adulterated Peppermint Oils

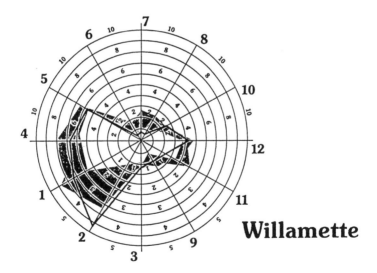

Figure 15.

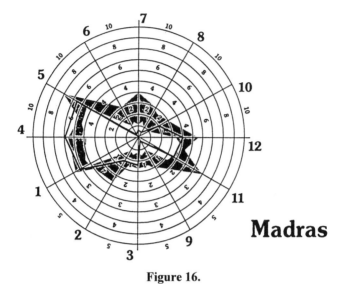

Madras

Figure 16.

Kennewick

Figure 17.

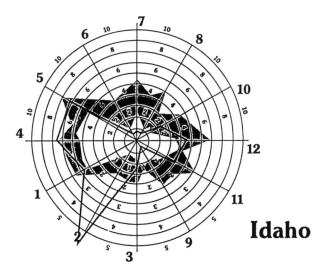

Figure 18.

Literature Cited

1. Lawrence, B. M.; Shu, C-K.; Harris, W. R. *Perfum. Flav.* **1989,** *14(6),* pp. 21-30.
2. Lawrence, B. M. In *Flavors and Fragrances: A World Perspective.* Lawrence, B. M.; Mookerjee, B. D.; Willis, B. J., Eds., Elsevier Sci. Publ., Amsterdam, 1988, pp. 161-169.
3. Lawrence, B. M. In *New Crops*; Janick, J.; Simon, J. E., Eds., J. Wiley & Sons, New York, 1993, pp. 620-627.
4. Lawrence, B. M. *Progress in Essential Oils.* **1992,** *17(5),* pp. 131-133.
5. Schmaus, G.; Kubeczka, K. H. In *Essential Oils and Aromatic Plants.* Baerheim Svendsen, A.; Scheffer, J. J. C., Eds., Martinus Nijhoff/Dr. W. Jank Publishers, Dordrecht, Netherlands, 1985 pp. 127-134.
6. Doré, D.; Jaubert, J-N, *Parfum. Cosmet. Arômes.* **1985,** *61*, pp. 79-85.
7. Lawrence, B. M. In *Advances in Labiatae Science;* Harley, R. M.; Reynolds, T., Eds.; p, Royal Botanic Gardens Kew, 1992, pp. 399-436.

Chapter 13

Component Analyses of Mixed Spices

C. K. Cheng[1], C. C. Chen[1], W. Y. Shu[2], L. L. Shih[1],
and H. H. Feng[1]

[1]Food Industry Research and Development Institute, P.O. Box 246,
Hsinchu 300, Taiwan, Republic of China
[2]Institute of Statistics, National Tsing-Hua University,
Hsinchu 300, Taiwan, Republic of China

A qualitative and quantitative technique for recognizing spices as well as the mixing ratios in an unknown spice blend is established via numerical analysis based on a database consisting of 355 different spices. Formulating the spice recognition problem as a constrained optimization problem is a key step. A similarity index measuring the degree of similarity between two spices is proposed, which is very useful in the process of spice recognition. For purposes of testing, an existing mixture (basil, 47.9%; clove, 12.8%; mint, 9.4%; sage, 23.8%; wintergreen, 6.1%) and a trial spice blend with known composition were analyzed by the technique. The results are surprisingly accurate.

Humans have been consuming spices and spice blends for several thousand years. Even to this day, spices and spice blends are highly valued commodities from all over the world (*1 - 2*).

From the flavor point of view, the major attributes of most spices are contributed by essential oils. These can be extracted by steam distillation (or similar process) of ground spices. With the advent of gas chromatography (GC) and the combined technique of GC/ mass spectrometry (MS), analyses and identifications of volatile components of essential oils extracted from different spices have been very thoroughly and well documented (*3-5*). However, most of the published reports were based on the analysis of an individual spice. When it comes to food processing, usually it is a spice blend instead of the individual spice which has been added to the formulation.

Detection of an individual spice in a spice blend can be achieved through sensory analysis, instrumental analysis, or the combination of both. When analyzed by a sensory method, which requires trained panelists, only qualitative results can be obtained (6). For semi-quantitative analysis, there have been two types of approaches to investigate the content of an individual spice in a spice blend. One is based on calculation of major components in a mixture as compared to published results. A good example has been demonstrated by Lawrence and Shu (4). The more advanced approach is based on a computer pattern recognition method. Chien (7) has shown a very genius mathematical method to analyze the presence of an individual essential oil in a mixture. In that report, four out of six essential oils with contents ranging from 5% (geranium, cedar wood, patchouli) to 15% (bergamot) were correctly identified. The other two oils, galbanum and Bourgeon de Cassis, were not picked up by the computer because of their low level (0.25%).

This paper presents a qualitative and quantitative analysis of an individual spice in a spice blend. The ratios of each spice were calculated by using a computerized numerical analysis with a data matrix composed of 355 spices and accumulated total of 922 volatile compounds. The principal part of this method is the solving of a constrained optimization problem.

Theory

The problem of recognizing the individual spice as well as the mixing ratios in a spice blend is first formulated as a constrained optimization problem. We then solve the problem by numerical methods. The procedure is described below.

Vector representation of spices. Each spice can be characterized by its components and composing ratios. Mathematically, each spice can be treated as a vector $X=(x_1,x_2,...,x_n)$, where x_i represents the ratio of the I-th compound in that specific spice. In the present study, 355 different kinds of spices were collected and 922 volatile compounds ($n=922$) were counted among these 355 spices. Therefore, n is equal to 922 and $x_1+x_2+...+x_{922}=1$. The 355 spices were represented by 355 vectors $X_1,X_2,...,X_{355}$. Using Anise #1 collected in the present study as an example, this is shown in Table I.

In Table I, each code number represents a single volatile compound identified in the essential oil, e.g., code no. 236 represents estragole, code no.

Table I . The volatile compounds of anise #1.

Code No. of compounds	Ratio (%)
156	2.513
187	1.487
236	4.975
250	0.121
292	0.312
293	0.221
295	0.010
454	2.382
531	0.784
532	0.030
540	86.683
666	0.221
667	0.261
Sum	100

540 represents trans-anethole. A spice blend X of individual spices contained in our collection can be viewed as a complex combination of vectors $X_1, X_2, ..., X_{355}$:

$$X = \beta_1 X_1 + \beta_2 X_2 + ... + \beta_{355} X_{355} \tag{1}$$

where β_i is the mixing ratio of the i-th spice in the spice blend . For a given unknown spice blend Y, identifying the spices contained in Y amounts to finding a convex combination $X = \beta_1 X_1 + \beta_2 X_2 + ... + \beta_{355} X_{355}$ as "close" to Y as possible. Methods for measuring the closeness between spice blends X and Y will be discussed next.

Similarity index For any two spice blends $X=(x_1, x_2, ...x_{922})$, and $Y=(y_1, y_2, ...y_{922})$, the similarity index, denoted $Index(X,Y)$, is defined as:

$$Index(X,Y) = \sum_{i=1}^{922} \sqrt{x_i} \sqrt{y_i}, \tag{2}$$

Basically, $Index(X,Y)$ has the following properties:
For any two spice blends X and Y,

1. $0 \leq Index(X,Y) \leq 1$ (3)

2. If $X=Y$, $Index(X,Y)=1$ (4)

3. If components in X can not be found in Y and vice versa, then $Index(X,Y)=0$.   (5)

Equations 3-5 indicate that if spice blend X is compared with itself, then *Index(X,X)*=1, if components in spice blend X can not be found in Y and vice versa, then *Index(X,Y)*=0, otherwise, the index shall be ranged between 1 and 0. Therefore, *Index(X,Y)* can be used to measure the similarity between X and Y.

Similarity index has its mathematical meaning. Any spice vector can be considered as a probability distribution on the set $\{1,2,....,922\}$. The well-known squared Hellinger distance between two probability distributions X and Y is defined as:

$$\sum_{i=1}^{922}\left(\sqrt{x_i}-\sqrt{y_i}\right)^2 = 2-2\cdot\sum_{i=1}^{922}\sqrt{x_i}\sqrt{y_i} = 2\left[1-Index(X,Y)\right],\tag{6}$$

It is very clear that the larger the *Index(X,Y)* is, the smaller the Hellinger distance between X and Y will be. Therefore *Index(X,Y)* directly reflects their similarity. For example, Anise#2 (*9*) and Anise #4(*10*) have many volatile compounds in common (Table II). The similarity index between these two spices is 0.964 indicating high degree of similarity. On the contrary, when volatile compounds of Anise #2 are compared with those of coriander #1 (*11*), as shown in Table III, an index value of 0 is obtained, indicating no similarity between these two spices.

Table II. Comparison of the volatile compositions of anise #2 and star anise #4. The similarity index = 0.964

Code No.	Ratio		Code No.	Ratio	
	Anise #2	Star anise #4		Anise #2	Star anise #4
20	0.00000	0.00081	436	0.00000	0.00081
40	0.00000	0.00092	437	0.00010	0.00581
60	0.00000	0.00102	440	0.00000	0.00112
156	0.00567	0.04563	450	0.00000	0.00051
236	0.00304	0.00346	454	0.00000	0.00112
257	0.00000	0.00509	467	0.00010	0.00051
292	0.00608	0.02954	480	0.00365	0.00428
293	0.00263	0.00428	499	0.00020	0.00346
303	0.00000	0.00092	518	0.00010	0.00000
331	0.00000	0.00061	531	0.00000	0.01039
335	0.00000	0.00122	532	0.00010	0.00183
375	0.00010	0.00051	540	0.97376	0.87044
386	0.00020	0.00387	552	0.00000	0.00143
388	0.00010	0.00000	666	0.00405	0.00000
425	0.00000	0.00041	745	0.00010	0.00000

Table III Comparison of the volatile compositions of anise #2 and coriander #1. The similarity index = 0.

Code No.	Ratio		Code No.	Ratio	
	Anise #2	Coriander #1		Anise #2	Coriander #1
6	0.00000	0.46519	211	0.00000	0.00505
7	0.00000	0.09284	231	0.00000	0.01615
8	0.00000	0.10394	236	0.00304	0.00000
11	0.00000	0.00303	267	0.00000	0.00303
13	0.00000	0.05853	292	0.00608	0.00000
14	0.00000	0.00706	293	0.00263	0.00000
15	0.00000	0.05651	348	0.00000	0.00202
32	0.00000	0.00303	349	0.00000	0.00202
50	0.00000	0.04339	353	0.00000	0.00505
58	0.00000	0.00202	357	0.00000	0.00101
61	0.00000	0.00303	375	0.00010	0.00000
62	0.00000	0.00101	386	0.00020	0.00000
64	0.00000	0.00202	388	0.00010	0.00000
65	0.00000	0.00202	390	0.00000	0.00706
66	0.00000	0.00101	399	0.00000	0.00101
67	0.00000	0.01413	402	0.00000	0.00505
71	0.00000	0.00404	437	0.00010	0.00000
83	0.00000	0.00706	467	0.00010	0.00000
87	0.00000	0.00605	480	0.00365	0.00000
91	0.00000	0.00908	499	0.00020	0.00000
125	0.00000	0.00101	518	0.00010	0.00000
143	0.00000	0.01312	532	0.00010	0.00000
144	0.00000	0.00807	540	0.97376	0.00000
145	0.00000	0.00101	666	0.00405	0.00000
156	0.00567	0.00000	745	0.00010	0.00000
210	0.00000	0.04440			

Numerical Analysis.   Given an unknown spice blend Y , we consider the spice recognition problem as that of finding the best ratio in which one can mix the 355 spices $X_1, X_2, ..., X_{355}$ so as to achieve the maximal similarity to the target spice blend Y . This problem is exactly equivalent to the following constrained optimization problem:

$$\begin{cases} maximize\ S(\underline{\beta}) \\ Subject\ to\ \beta_1 + \beta_2 + ... + \beta_{355} = 1\ ,\ \beta_i \geq 0, \end{cases} \tag{7}$$

where $\underline{\beta} = (\beta_1, \beta_2, ..., \beta_{355})$ represents a set of mixing ratios and the objective function S is defined as:

$$S(\underline{\beta}) = Index(\beta_1 X_1 + \beta_2 X_2 + ... + \beta_{355} X_{355},\ Y) \tag{8}$$

To determine the best ratio $\underline{\beta}^*$ so as to achieve the maximal value of $S(\underline{\beta})$, we employ optimization   theory and numerical   methods (see Luenberger (8)) and computer programs are designed accordingly. The nonzero components of $\underline{\beta}^*$ indicate the presence of the corresponding spices.

Experimental Section

Preparation of mixed spices for testing.   A testing mixture of spice blend was prepared by   mixing powdered spices purchased from local trading company. The formulation is shown   in Table IV .

Table IV. The formulation of mixed spices for testing.

Spices		Wt(g)
Black pepper		27.00
Allspice		27.02
Celery seed		13.52
Clove		1.62
Garlic		27.01
Star anise		61.03
	Sum	157.20

Isolation of volatile components from spices.   One hundred grams of ground spice blend   shown in Table IV were added to a three-neck bottle containing 300ml of distilled water. This slurry was steam distilled and extracted   with dichloromethane (Fisher Scientific) for 2 hours in an apparatus similar to that described by Likens-Nickerson(12). The organic layer was separated, dried over anhydrous sodium sulfate (E. Merck), evaporated to minimal volume in a Vigreux column, and then concentrated to about 0.5 ml under a gentle stream of nitrogen.

GC. GC analyses were carried out on a Varian 3400 chromatograph equipped with a 30 m x 0.25 mm i.d. fused silica capillary column (DB-WAX, J&W Scientific, USA). The linear flow rate of the carrier gas (H_2) was 43 cm/s. The oven temperature was programmed from 40°C to 200°C at a rate of 2°C/min with initial holding at 40°C for 2 min. The injector and detector (FID) temperatures were set at 200°C and 210°C, respectively.

GC/MS. GC/MS was carried out on a Finnigan Mat ITD. GC conditions were the same as those mentioned above. Identification of volatile compounds was based on comparison of retention index and mass spectra with those of authentic compounds.

Set up of the data matrix. The raw data were collected from published literature such as those cited by TNO-CIVO (*3*), ESO (*5*), or those listed in the present study. Take essential oil of black pepper (*13*) as an example, the original data format is shown in Table V, the volatile compounds were then arranged into coded number as those of Anise #1 shown in Table I. So far, a 355 (spices) by 922 (volatile compounds) data matrix has been established. Of the 355 spices, there exists repeated collection of data from the same spice from different geological areas or different publications. For examples, 4 cloves, 5 star anises, 9 anises, 6 corianders, 8 allspices, 2 black peppers and 27 basils, are included in the present data matrix.

Results and Discussion

Test run of known spice blends from published literature. The spice blends which contain basil, cinnamon leaf, peppermint, sage and wintergreen as reported by Lawrence and Shu (*4*) were used to test the effectiveness of the present theory. The ratios of individual spice are shown in Table VI, the coded numbers of identified volatile compounds in the spice blends are shown in Table VII. Computer output of numerical analyses of compounds shown in Table VII are listed in Table VIII, the similarity index of the testing result is 0.959, indicating a high level of confidence. In Table VIII, there are repeated identifications of the same type of spice, e.g., the appearance of basils #3, #19, #24, #26 and #27; cloves #1 and # 4; and mentha #1 and # 12. In fact, the repeated appearance of the same spice indicates the closeness of these spices. It was also confirmed that the similarity index within the same spice listed in Table VIII usually ranged from 1.0 to 0.90 (data not shown). In order to

Table V. Volatile compounds of essential oil of black pepper

Compounds	%
Sabinene	22.50
alpha-Pinene	17.50
Limonene	17.00
delta-3-Carene	13.50
beta-Caryophyllene	5.00
alpha-Humulene	1.00
Carvone	0.10
Eugenol	0.10
Linalool	0.10
Myristicin	0.10
Nerolidol (unknown isomer)	0.10
Piperonal	0.10
cis-Sabinene hydrate	0.10
trans-Sabinene hydrate	0.10
Safrole	0.10
Terpinen-4-ol	0.10
alpha-Terpineol	0.10
Total 17 components	77.60

(Source: Adapted from ref . *13*)

Table VI . Reported composition of testing mixture of essential oils

Essential oils	%
Exotic basil	42.1
Cinnamon leaf	15.8
Peppermint	15.8
Sage	21.1
Wintergreen	5.2

(Source: Adapted from ref . *4*)

Table VII. Volatile compounds identified from Table VI.

Code No.	Compound	%
437	alpha-Pinene	1.3
182	Camphene	1.5
467	beta-Pinene	0.7
335	Myrcene	0.5
292	Limonene	1.1
60	1.8-Cineole	4.8
444	alpha-Thujone	5.6
470	beta-Thujone	1.6
185	Camphor	4.7
297	Menthone	3.0
762	Menthofuran	0.4
293	Linalool	0.9
296	Menthol	4.8
236	Methyl chavicol	34.7
177	Borneol	1.7
325	Methyl salicylate	4.5
197	Cinnamaldehyde	0.1
250	Eugenol	9.4
251	Eugenyl acetate	0.4

(Source: Adapted from ref. 4)

Table VIII. Ratios of essential oils as analyzed by numerical analyses, the similarity index =0.959.

Spice #	%	Assignment
Basil #3	9.7393	
Basil #19	2.6187	
Basil #24	2.8070	
Basil #27	12.8562	
Basil #26	18.7463	Basil #26
Bay leaf #1	2.7733	
Cinnamon #4	0.4658	
Clove #1	0.5672	
Clove #4	10.6274	Clove #4
Cornmint #8	7.4766	Cornmint #8
Mentha #1	0.5556	
Mentha #12	2.6013	
Rosemary #3	2.7350	
Sage #2	19.4858	Sage #2
Wintergreen #1	5.9446	Wintergreen #1

simplify the result, the ratios of those spices and herbs which appeared repeatedly were summed up and assigned to the spice with the highest ratio, as shown in the third column of Table VIII. Table IX shows the summed up ratios of spices from Table VIII, as compared with those proposed by Lawrence and Shu (4). In Table IX, clove #4 and cornmint #8 were not reported in the previous study (4), instead, cinnamon leaf and peppermint, respectively, were assigned. When the similarity index between these two seemingly different pairs of spices (clove #4 vs. cinnamon leaf; cornmint #8 vs. peppermint) were compared, an index value of 0.95 was observed in the former, indicating a high degree of closeness (similar in composition of essential oil), the latter pair showed a less satisfactory index value of 0.83, still this was a good match with acceptable confidence. It should be considered that the results in Table IX were from data in reference 4 that was used in the data matrix of the present study. The ratios of spices analyzed by the numerical method were very accurate when compared to the original data. This is a good demonstration that the approach used in the present study can be used to examine the ratios of any spice blend, as long as the individual spice is included in the data matrix (355 spices).

Table IX. Comparison of the numerical method with previous publications.

Spices assigned	% [a](cal.)	Spices[b]	%
Basil #26	47.9144	Exotic basil	42.1
Clove #4 [c]	12.7725	Cinnamon leaf	15.8
Cornmint #8 [d]	9.3919	Peppermint	15.8
Sage #2	23.8251	Sage	21.1
Wintergreen #1	6.0962	Wintergreen	5.2

a:summed up values from Table VIII.; b: data from ref 4.; c:*Index* (Clove #4, Cinnamon Leaf #4)=*0.95;* d:*Index* (Cornmint #8 , Pepper #8)=*0.83*

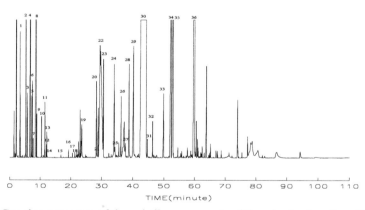

Fig 1. Gas chromatogram of the volatile components of the mixed spices from Table X.

Analysis of testing spice blend prepared in the laboratory. The formulation of a real spice blend used in the present study is shown in Table IV. Volatile compounds from this spice blend were isolated by simultaneous steam distillation and solvent extraction. Identification of volatile compounds was accomplished by comparing the retention index and mass spectra with those of authentic samples and published publications (*14 - 15*). Figure 1 shows the gas chromatogram of isolated volatile compounds from the testing spice blend. Percent ratios of these volatile compounds identified are shown in Table X. Compounds in Table X were then coded by number in the same manner as those of Table VII. The original results of numerical analyses are shown in Table XI, similar in format to those of Table VIII. Recognition of spices in this spice blend showed a very high similarity index (0.981), indicating high level of confidence. Those spices that appeared repeatedly with a high similarity index were assigned to the spice with highest ratio, i.e. the ratios of anise and star anise were combined, as mentioned earlier, that is because both anise and star anise listed are similar in composition. The same holds true for the combinations of black pepper and allspice. The only spice which is determined by numerical analysis but not added into the testing mixture is marjoram (ca. 0.43%).

The comparisons of ratios of spices as determined by numerical method and those calculated from the content of essential oil are shown in Table XII. Sensory analysis of a 0.4% solution of the actual spice blend as compared with that of a predicted spice blend have shown little difference. The accuracy of this test is surprisingly good for star anise, black pepper and garlic. The ratios of allspice and clove are quite good, although some deviations from the actual data exists. Celery has been correctly identified but the ratio determined in the present test (only 0.03 %) is quite low, probably caused by technical difficulty of our MS database in assigning the correct volatile compounds of celery. As to the appearance of marjoram, although the ratio assigned to this spice is small (0.43%), the similarity index between marjoram and other identified spices has not shown any high degree of similarity. Probably this deviation is caused by the low value assigned to celery (0.03% vs. 1.78%).

Conclusion

The present study has shown that the ratios of individual spices in a spice blend can be accurately solved by a numerical method when compared with previous reports (*4, 7*). Spices in a spice blend with an essential oil content as low as 0.37% can be

Table X. Identified volatile compounds from the spice blends from Table IV

Peak No [a]	%	Compounds
1	0.29	alpha-pinene
2	0.67	beta-pinene
3	0.19	sabinene
4	1.50	delta-3-carene
5	0.16	alpha-phellandrene
6	0.21	myrcene
7	0.05	alpha-terpinene
8	2.28	limonene
9	0.10	1,8-cineole
10	0.12	gamma-terpinene
11	0.18	para-cymene
12	0.04	allyl methyl disulfide
13	0.07	terpinolene
14	0.01	methyl trans-1-propenyl disulfide
15	0.01	dimethyl trisulfide
16	0.03	pentylbenzene
17	0.02	4-isopropenyl-1-methylbenzene
18	0.02	linalool oxide
19	0.25	alpha-copaene
20	0.34	linalool
21	0.02	trans-alpha-farnesene
22	3.37	beta-caryophyllene
23	0.49	terpinen-4-ol
24	0.45	delta-cadinene
25	0.08	estragole
26	0.25	alpha-terpineol
27	0.09	carvone
28	0.45	cis-anethole
29	0.90	diallyl trisulfide
30	70.16	trans-anethole
31	0.06	para-cymen-8-ol
32	0.14	3-methyl-1-phenyl-1-butanone
33	0.39	beta-caryophyllene oxide
34	3.02	anisaldehyde
35	2.17	methyleugenol
36	11.46	eugenol
Total	100.00	

a: Number refers to Fig 1.

Table XI. Computer output of ratios of essential oils as
analyzed by numerical analyses. The similarity index = 0.981

Spice #	Calculated %	Assignment
Anise #1	3.9245	
Anise #2	17.8341	
Anise #4	9.5149	
Anise #5	7.0813	
Anise #6	9.6366	
Anise #7	0.7293	
Star anise #2	9.9241	
Star anise #4	19.3589	Star anise #4
Celery #5	0.0305	Celery #5
Clove #4	4.5803	Clove #4
Garlic #1	0.4170	garlic #1
Marjoram #2	0.4275	marjoram #2
Black pepper #1	0.2784	
Black pepper #2	5.7291	Black pepper #2
Allspice #4	0.3164	
Allspice #5	1.9528	
Allspice #8	3.7578	Allspice #8

Table XII. Ratios of individual spice in the spice blends after adjustment
according to closeness of similarity index

spices	wt (g)[a]	essential oil content (%)[b]	% oil in the mixture (cal)[c]	% oil in the mixture (exp)[d]
black pepper	27.00	1.76	6.43	6.29
allspice	27.02	2.58	9.44	6.31
celery	13.52	0.97	1.78	0.03
clove	1.62	7.67	1.68	4.80
star anise	61.03	9.72	80.31	81.70
garlic	27.01	0.10	0.37	0.44
marjoram				0.43

a:Data from Table IV.;b: Data obtained from this study.;c:Normalized per cent
distribution; d:Ratios combined from Table XI.

accurately determined without significant deviation from the actual value. It is the utmost hope of the authors that application of this technique can definitively shorten the time and effort used to determine the ratio of individual spice in a spice blends.

Acknowledgements

This work was supported by the Ministry of Economic Affairs, Taipei, Taiwan, Republic of China. We are also grateful to our laboratory staffs for technical assistance.

Literature Cited

1.Farrell,K.T. In *Spices,Condiments,and Seasonings*. Van Nostrand Reinhold, New York.**1990.**

2.Giese, J. *Food Technol.,* **1994.** 48(4),88.

3.*Volatile Compounds in Foods Maarse*, H., Visscher, C. A., Eds., TNO-CIVO, Zeist, the Netherlands **1989**, Vol. 1-2.

4.Lawrence, B.; Shu, C. K. In *Flavor Measurement.* Ho, C.T.; Manley, C.H., Eds.; Marcel Dekker, New York, **1993,** pp.267-328..

5. *ESO: The Complete Database of Essential Oils*. BACIS, the Netherlands, **1995**.

6.Salzer,U.J. *CRC Crit. Rev. Food Sci. Nutri.,* **1977,** 9,345.

7.Chien, M. *Anal. Chem.* **1985,** 57, 348.

8.Luenberger,D.G. *Optimization by Vector Space Methods.* John Wiley & Sons, New York.**1968.**

9.Tabacchi, R.; Garnero, J.; Buil, R. *Rivista Ital.,* **1974,** 56, 683.

10.Cu, J.Q. In *10th Int. Congress Ess. Oils, Fragr.& Flav.,* Washington, D.C., **1986,** pp.231-241.

11.Potter, T.L.; Fagerson, I.S. *J. Agric. Food Chem.,* **1990,** 38, 2054.

12. Likens, S. T., Nickerson, G. B., *Am. Soc. Brewing Chemists Proc.* **1964,** 5.

13.Wrolstadt, R.E.; Richard, H.M.; Jennings, W.G. *J. Food Sci.,***1971,** 36,584.

14. Jennings, W.; Shibamoto, T. In *Qualitative Analysis of Flavor and Fragrance Volatiles by Glass Capillary Gas Chromatography.* Academic Press, New York. **1980.**

15. *Monoterpenes: Infrared, Mass, H-NMR, and C13 NMR Spectra and Kovats Indices.* Swigar, A. A.; Silverstein, R. M. Eds.; Aldrich Chem. Co. Inc., Milwaukee, Wisconsin, **1981.**

ANTIOXIDANT PROPERTIES

Chapter 14

Antioxidative Activity of Spices and Spice Extracts

Helle Lindberg Madsen, Grete Bertelsen, and Leif H. Skibsted

Department of Dairy and Food Science, The Royal Veterinary
and Agricultural University, Rolighedsvej 30,
DK–1958 Frederiksberg C, Denmark

The antioxidative activity of spices and spice extracts can usually be traced
back to their content of phenolic compounds. Plant phenols may scavenge
free radicals involved in lipid peroxidation as has been documented in
several model systems, although other mechanisms should be considered
especially in relation to the early stages of oxidative deterioration. Phenolic
compounds isolated from spices have been found to react with hydroxyl
radicals with nearly diffusion controlled reaction rates. An assay based on
a combination of determination of phenol equivalents and determination of
radical scavenging capacity by the ESR spin trapping technique confirms
the nearly diffusion controlled reaction rates and may prove useful for
exploration of new plant materials and for adjustment of extraction
procedures, including selection of solvent. This and other assays based on
oxygen depletion measurements are recommended prior to final testing in
real foods. Co-extracted chlorophylls in spice extract present a problem as
photosensitizers in food exposed to light during storage and use.

Addition of spices to food is an established procedure in most cultures. The seasoning
contributes a pleasant flavour, and at the same time a number of compounds possessing
antioxidative activity are added. Scientific investigation of the antioxidative activity of
spices has been initiated during the last few decades and an understanding of the
mechanism of the antioxidative activities is emerging. Early work by Chipault and
coworkers (1-3) in the fifties documented significant reduction in oxidation in lard and
edible oils. The most pronounced effects were found for rosemary and sage, observations which have been verified in most later investigations (4-6). However, a great
variety of other spices investigated show similar antioxidative activity and notably their
relative efficiencies are strongly dependent on the actual food.

Oxidation and Antioxidants

During oxidation free radicals are generated continuously in the propagation phase. In this phase a lipid radical ($R\cdot$) reacts very quickly with (triplet) oxygen in a reaction controlled by diffusion. The formed peroxyl radical ($ROO\cdot$) can abstract a hydrogen from an unsaturated lipid molecule ($R'H$). This reaction leads to generation of a new lipid radical ($R'\cdot$) and continued activity in the chain reaction. The hydroperoxides ($ROOH$), the products formed in the chain reaction, are without smell and taste. Hydroperoxides are easily decomposed to yield alkoxyl radicals ($RO\cdot$). These radicals can further abstract a hydrogen from a lipid molecule in effect starting a new chain reaction contributing further to the propagation phase (*7*) (Figure 1). Reaction products from the degradation of the initial formed hydroperoxides, the secondary oxidation products, are carbonyl and carboxyl compounds which give rise to the unpleasant taste and smell of oxidized food (*8*).

An efficient way to delay oxidation is scavenging by antioxidants of the free radicals generated in the propagation phase or during the break down of the hydroperoxides, *i.e.* scavenging of either the peroxyl radicals or the alkoxyl radicals. The critical level needed of such primary antioxidants to be effective in a given product corresponds to the concentration necessary to inhibit all chain reactions started by the initiation process. As long as the concentration of the antioxidants is above this critical concentration, the total number of radicals is kept at a constant low level, a time period which is defined as the induction period. During the induction period the antioxidant is gradually depleted and when the critical concentration is reached, radicals will escape from reaction with the antioxidant, and the concentration of hydroperoxides will increase. The high level of hydroperoxides will further increase the concentration of radicals, and the remaining antioxidant will be used up completely (*9*). With all the antioxidants consumed, the oxidative processes will accelerate, and the increase in the production of secondary oxidation products will lead to a progressing deterioration of the product.

Phenolic compounds act as primary antioxidants by donating a hydrogen atom to either the alkoxyl radicals (equation 1) or the peroxyl radicals (equation 2) in irreversible reactions. The reactions lead to generation of an antioxidant free radical ($A\cdot$) with a lower energy compared to the energy of $RO\cdot$ and $ROO\cdot$, *i.e.* the reaction is enthalpy driven.

$$AH + RO\cdot \ \rightarrow \quad A\cdot + ROH \tag{1}$$
$$AH + ROO\cdot \rightarrow \quad A\cdot + ROOH \tag{2}$$

The relatively high stability of $A\cdot$ reduces its ability to abstract a hydrogen from a lipid molecule in effect breaking the chain reaction. The efficiency of phenolic compounds as scavengers of free radicals is due to the high stability of the generated phenoxy radical, which is caused by delocalisation of the unpaired electron in the aromatic ring.

However, phenols are not active as antioxidants unless substitution at either the ortho or para position has increased the electron density at the hydroxy group and lowered the oxygen-hydrogen bond energy, in effect increasing the reactivity towards the lipid free radicals. Substitution in phenolic compounds at the meta position has a rather limited effect, and compounds like resorcinol with a hydroxyl group in the meta

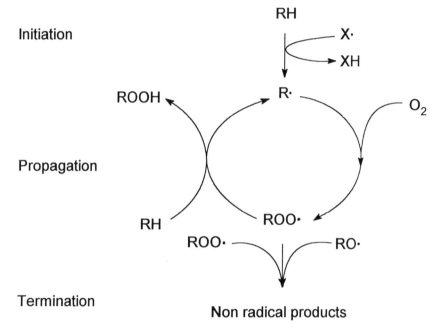

Figure 1. Lipid oxidation with an initiation phase, propagation phase and termination phase.

position are thus poor scavengers of free radicals compared to similar compounds with substitution at the ortho or para position. The presence of additional hydroxy groups, either at the ortho or para position, further increases the antioxidative activity of the compound as intra-molecular hydrogen bonds stabilize the phenoxyl radicals (9).

Antioxidative Compounds in Spices

Phenolic compounds constitute the largest proportion of know natural antioxidants. It should, however, be noted that only a minor part of the compounds in spices which have antioxidative activity has been isolated and identified. Nakatani and coworkers (10,11) have reported the structure and characterized the antioxidative properties of several phenolic diterpenes isolated from rosemary (*Rosmarinus officinalis* L.). From the same plant, compounds like carnosic acid, carnosol (12-14), rosmaridiphenol and rosmari-quinone (15,16) have been identified (Figure 2) as antioxidants. Moreover a number of the compounds found in rosemary have been found in the botanically closely related sage (*Salvia officinalis* L.) (12,13) and summer savory (*Satureja hortensis* L.) (17,18).

In other spices, flavonoids alone or together with other phenolic compounds have been found to contribute to the antioxidative activity. In oregano (*Origanum vulgare* L.) various antioxidative compounds have been isolated, and among the active components four flavonoids were identified (19). From thyme (*Thymus vulgaris* L.), which like oregano also is a member of the *Labiatae* family, antioxidative compounds have been isolated and identified as dimers of thymol and flavonoids (20). The antioxidative activity of pepper (*Piper nigrum* L.) can, at least partially, be ascribed to the presence of glycosides of the flavonoids kaempferol, rhamnetin and quercetin and at least five different phenolic amides (21,22).

A variety of different volatile compounds such as the terpenoids thymol, carvacrol, eugenol, carvone and thujone, which are character-impact compounds for important spices (23,24), have antioxidative activity but the use of these compounds as antioxidants for different foods are limited by the characteristic flavour of the particular compound.

Evaluation of the Radical Scavenging Activity

Screening of antioxidative activity in various model systems is important prior to testing or application of antioxidants in foods. Such model systems are more rapid compared to food storage experiments, and the model systems might even be more informative in relation to antioxidant mechanisms.

Electron spin resonance (ESR) spectroscopy is currently being introduced to different areas of application in food science. The technique may provide valuable information concerning the elementary processes in lipid oxidation and the nature of reaction intermediates. A number of antioxidants including extracts or compounds isolated from spices have been investigated by different ESR technique. The antioxidative activity of the fraction containing the essential oils from oregano (*Origanum vulgare* L.), summer savory (*Satureja hortensis* L.) and thyme (*Thymus vulgaris* L.) has been documented in a system where oxidative stress and free radical reactions were induced either by UV irradiation or the superoxide radical. Free radicals can be detected directly

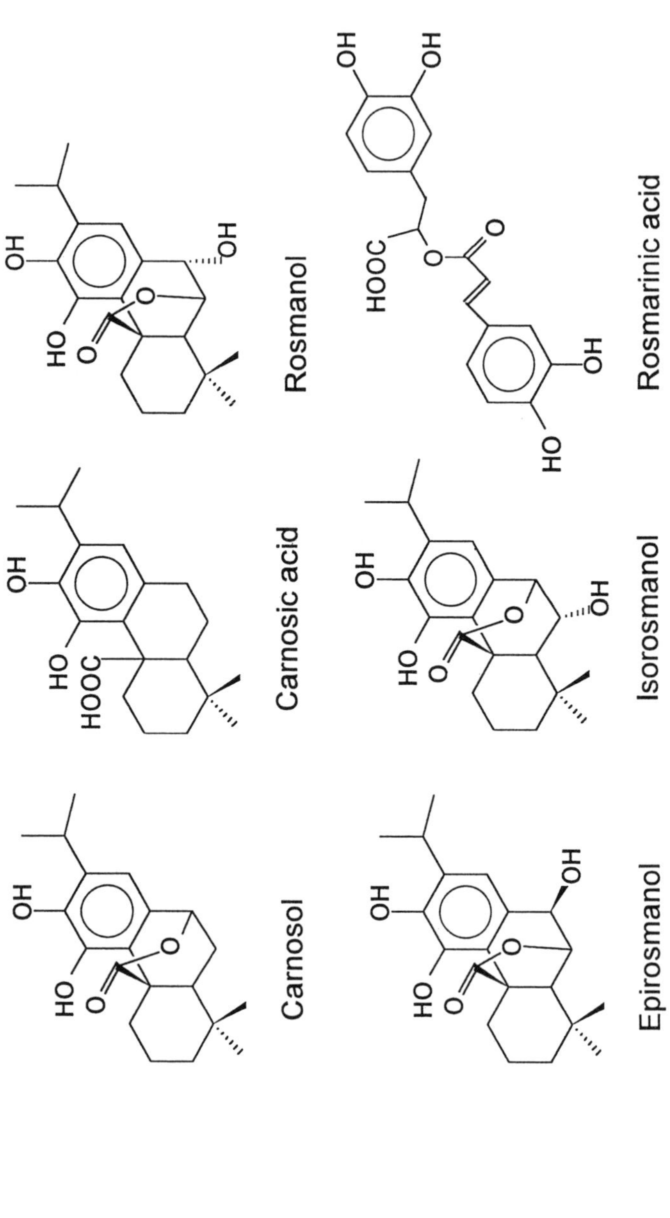

Rosmanol

Rosmarinic acid

Carnosic acid

Isorosmanol

Carnosol

Epirosmanol

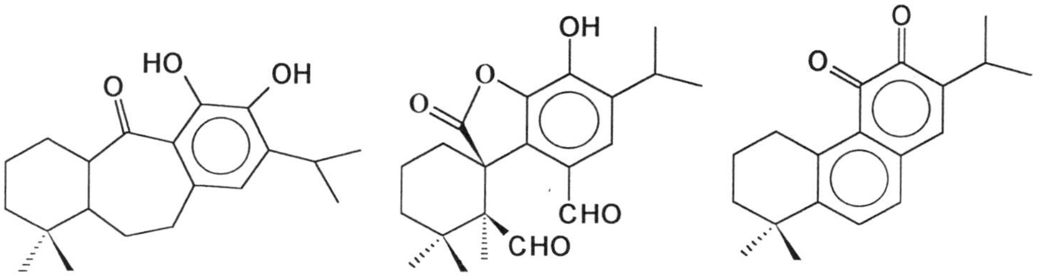

Rosmaridiphenol Rosmadial Rosmariquinone

Figure 2. Chemical structure for compounds isolated from rosemary *(Rosmarinus officinalis* L.).

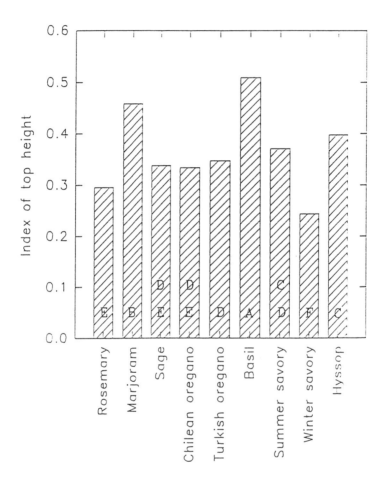

Figure 3. The hydroxyl radical scavenging activity of nine different spices all belonging to the family *Labiatae*. The activity was measured by using electron spin resonance technique, and the indexes were calculated on the basis of top height from data of reference (*28*). A low value of the index indicates an efficient scavenger. Columns with different letters are significantly different at 5% level.

by ESR spectroscopy, and have in the spice extract of oregano, thyme and summer savory been identified as free radicals derived from thymol and carvacrol. Notably, thymol and carvacrol radicals were also present in untreated essential oils indicating the high stability of the radicals and the potential of the compounds as scavengers of oxygen-centred free radicals (25). Experiments using spin trapping technique are based on competition between the compound expected to have antioxidative properties and a "spin trap" in reaction with free radicals. The *in vitro* antioxidant properties of rebamipide, a novel antiulcer agent, have thus been documented by the spin trapping technique (26), and the scavenging activity of carnosine and related dipeptides have been further investigated (27). In an investigation with focus on the early stages of oxidation in foods, the ability of spice extracts to inhibit the initiation of free radical process was evaluated by adapting the ESR spin trapping technique (28). The extracts of the spices obtained by ethanol/water extraction were freeze dried and redissolved in water. The water soluble compounds, partly phenols, were found to reduce the signal intensity in the ESR spectrum of the spin trap/free radical adduct when hydroxyl radicals were generated by the Fenton reaction, indicating a high scavenging potential of the spice extracts against hydroxyl radicals. Extracts of each of nine different spices investigated, all belonging to the *Labiatae* family, were capable of scavenging hydroxyl radicals (Figure 3). Most efficient was winter savory, rosemary, sage and the two oregano species. Kinetic studies gave an estimate of the rate constant between hydroxyl radicals and phenols in the extract of Turkish oregano. The value of the second order rate constant was slightly above that for a diffusion controlled bimolecular reaction in water, and the high value indicated that compounds other than phenols were involved in the competition with the spin trap in the reaction for the hydroxyl radicals. Rate constants for reactions between compounds isolated from spices and different radicals have, however, only been determined for a few substances and further studies should be encouraged. An estimate of the rate constant of $4.8 \cdot 10^{10}$ M $^{-1}$ s $^{-1}$ between hydroxyl radicals and eugenol is, however, an example (29). This corresponds to a very fast reaction with a rate constant similar to the rate constant found for the reaction between hydroxyl radicals and phenols in extracts of Turkish oregano (28). These results together document that hydroxyl radicals are very reactive, and model systems based on other radicals should be developed for a differentiation between different groups of free radical scavengers. The second-order rate constant determined for the superoxide scavenging activity of eugenol (a volatile present in clove, cinnamon and basil (29)) has been used for comparison of phenolic antioxidants and the enzyme superoxide dismutase. Superoxide dismutase was most effective in deactivating superoxide, intermediates rate constants were found for both eugenol and guaiacol, while the activity of phenol was rather poor (30).

At later stages in the oxidation process the scavenging activity of antioxidants towards peroxyl radicals is of relevance. Aruoma et al. (31) investigated the rate of reaction of a number of antioxidants with the trichloromethyl peroxyl radical. Carnosic acid had a rate constant of $2.7 \cdot 10^7$ M^{-1} s^{-1}, about 10 times higher than the rate constant of carnosol. The reactivity of carnosic acid with the peroxyl radicals was comparable to propyl gallate, and it could be concluded that carnosic acid is effective as a scavenger of trichloromethyl peroxyl radicals, although the reactivity of carnosic acid was not as high as the reactivity of ascorbic acid and the water soluble tocopherol analogue, trolox C. Determination of the rate constants for the reaction between hydroxyl radicals and either carnosic acid or carnosol again showed rate constants which indicated that the reactions were essentially diffusion controlled (31).

Antioxidative activity in model systems

While determination of rate constants for scavenging of free radicals by antioxidants is highly relevant for establishment of structure/activity relationships, it is essential for practical food application to construct heterogeneous model systems with phase transfer. Frankel et al. (32) evaluated rosmarinic acid, carnosic acid and carnosol in such model systems by measuring both primary and secondary oxidation products and showed that the antioxidative activity depends strongly upon the substrate. The compounds containing a carboxy group (rosmarinic acid and carnosic acid) were the most effective antioxidants in bulk corn oil whereas the antioxidative activity of carnosol was only limited in this oil. However, in an o/w emulsion a high antioxidative activity was seen for carnosol and carnosic acid while only slight antioxidative or even prooxidative activity was found for rosmarinic acid. The observation that the more polar antioxidants are more active in pure lipids, and the non polar antioxidants most active in a polar substrate, and for which the term "polar paradox" has been introduced, confirms previous findings for tocopherols and ascorbic acid derivatives as antioxidants (33). In the bulk oil the hydrophilic antioxidants are oriented in the oil/air interface providing optimal protection of the lipids against oxygen radicals while the hydrophobic antioxidants dissolves in the homogeneous lipid phase. The opposite situation with hydrophobic antioxidant concentration in the oil/water interface is encountered in the emulsion system, where the hydrophobic antioxidants, like carnosol, are most efficient. These observations may at least partially explain the variation of antioxidative activity seen for different spices in different foods (3), and the need for investigations in such heterogenous model systems and in the actual food products prior to practical use is obvious.

Antioxidative activity in food

The results obtained by Frankel et al. (32) using food models might explain the qualitative difference observed between antioxidative activity of summer savory and rosemary in a meat storage experiment and in a storage experiment with dressing. In precooked pork meat balls no significant difference between the antioxidative activity of rosemary and summer savory was seen, and both spices were able to retard the development of oxidative off-flavour, measured by determination of thiobarbituric acid reactive substances (Madsen et al. The Royal Veterinary and Agricultural University, Frederiksberg, DK, unpublished data). These results were, however, different from those found for a dressing with a high fat content (50%). In the dressing the addition of rosemary was found to be significantly more effective in protection against oxidation compared to summer savory (Madsen et al. The Royal Veterinary and Agricultural University, Frederiksberg, DK, unpublished data). Although the nature and quantity of the active compounds present in the two spices were not determined, it may be speculated that differences in the hydrophilic or hydrophobic properties of the antioxidants may be important. It is well-known, that rosmarinic acid is an important antioxidative component in summer savory (*18*). The low efficiency of this substance in bulk oil might provide a rationale for the reduced antioxidative activity observed for summer savory in the dressing. Other factors such as differences in pH in the two foods may also contribute to the different activities. Frankel *et al.* (*32*) have also studied the antioxidative activity of carnosol, carnosic acid and rosmarinic acid in emulsions with different pH. At pH 4 and pH 5, where the carboxyl group is protonated to a significant extent, the antioxidative activity of carnosol and carnosic acid were much higher than the activity at pH 7 where the carboxylate form dominates. This pH dependency might be explained by the change in polarity and oxidation potential or by a higher stability of the compounds at lower pH. An increased knowledge of the effects of pH will be beneficial for a more rational use of plant antioxidants.

Further perspectives and developments

A three-step procedure as the one outlined encompassing (i) determination of radical scavenging activity of spice extract, (ii) testing in model systems, and (iii) final application in foods may prove useful for future work on natural antioxidants. Different solvents should also be explored for extraction of spices. Different polarity of the solvents might be used for "tuning" of extracts to different products of different hydrophobic/hydrophilic balance. A higher antioxidative activity was found of a hexane extract of rosemary and sage compared to other solvents by using successive extraction of solvents of increasing polarity (*34*). However, most investigations report a high antioxidative activity in extracts obtained from polar solvents (*1,35,36*). The use of large amounts of organic solvents is in general unwanted for environmental resonance, so alternative extraction methods should be considered. The use of supercritical CO_2 extraction, which has already shown positive results (*37*), might be getting more attention since the technique is efficient and also protects the phenolic compounds against degradation during extraction.

The interaction between different antioxidants such as tocopherols, ascorbic acid and spice components and possible synergisms should also be explored. Fang & Wada (*38,39*) already investigated the synergism between α-tocopherol and extract of rosemary in ethanol. The results support the hypothesis that rosemary antioxidants regenerate oxidized α-tocopherol. Similarly Ternes & Schwarz (*40*) found an increased stability of tocopherols when rosemary extract was added to meat samples. To obtain more information about the antioxidative compounds a determination of the reduction potential of key components (*6,41*) would be of interest for ranking of antioxidants in order to determine which of the spice antioxidants can regenerate the tocopherols. The problems encountered with photosensitizing chlorophylls co-extraction from the spices making the food product sensitive to light (Madsen *et al.* Royal Veterinary and Agricultural University, Frederiksberg, DK, unpublished data) also needs further investigation.

Acknowledgments

This research was part of the frame program "Natural antioxidants from plants" sponsored by the FØTEK-program through National Food Agency and LMC-Centre for Advanced Food Studies.

Literature Cited

1. Chipault, J.R.; Mizuno, G.R.; Hawkins, J.M.; Lundberg, W.O. *Food Res.* **1952**, *17*, pp. 46-55.
2. Chipault, J.R.; Mizuno, G.R.; Lundberg, W.O. *Food Res.* **1955**, *20*, pp. 443-448.
3. Chipault, J.R.; Mizuno, G.R.; Lundberg, W.O. *Food Technol.* **1956**, *10*, pp. (5) 209-211.
4. Palitzsch, A.; Schulze, H.; Metzl, F.; Baas, H. *Fleischwirtsch.* **1969**, *49*, pp. 1349-1353.
5. Gerhardt, U.; Schröter, A. *Gordian* **1983**, *9*, pp. 171-176.
6. Gerhardt, U.; Böhm, T. *Fleischwirtsch.* **1980**, *60*, pp. 1523-1526.
7. Kanner, J.; German, J.B.; Kinsella, J.E. *CRC Crit. Rev. Food Sci. Nutr.* **1987**, *25*, pp. 317-364.
8. Shahidi, F. In *Lipid oxidation in food;.* St.Angelo, A.J. Ed. American Chemical Society:.Washington DC, **1992**, pp. 161-182.
9. Pokorný, J. In *Autoxidaton of unsaturated lipids;.* Chan, H.W., -S. Ed. Academic Press:.London, **1987**, pp. 141-206.
10. Nakatani, N.; Inatani, R. *Agric. Biol. Chem.* **1981**, *45*, pp. 2385-2386.
11. Nakatani, N.; Inatani, R. *Agric. Biol. Chem.* **1984**, *48*, pp. 2081-2085.
12. Brieskorn, C.H.; Fuchs, A.; Bredenberg, J.B.; McChesney, J.D.; Wenkert, E. J. *Org. Chem.* **1964**, *29*, pp. 2293-2298.
13. Brieskorn, C.H.; Dömling, H.-J. *Z. Lebensm. -Unters. -Forsch.* **1970**, *141*, pp. 10-16.
14. Wu, J.W.; Lee, M.-H.; Chang, S. *J. Am. Oil Chem. Soc.* **1982**, *59*, pp. 339-345.
15. Houlihan, C.M.; Ho, C.-T.; Chang, S.S. *J. Am. Oil Chem. Soc.* **1984**, *61*, pp. 1036-1039.
16. Houlihan, C.M.; Ho, C.-T.; Chang, S.S. *J. Am. Oil Chem. Soc.* **1985**, *62*, pp. 96-98.
17. Reschke, A. *Z. Lebensm. -Unters. -Forsch.* **1983**, *176*, pp. 116-119.
18. Bertelsen, G.; Christophersen, C.; Nielsen, P.H.; Madsen, H.L.; Stadel, P. *J. Agric. Food Chem.* **1995**, *43*, pp. 1272-1275.
19. Vekiari, S.A.; Oreopoulou, V.; Tzia, C.; Thomopoulos, C.D. *J. Am. Oil Chem. Soc.* **1993**, *70*, pp. 483-487.
20. Nakatani, N. In *Phenolic compounds in food and their effects on health II;.* Huang, M.-T., Ho, C.-T., Lee, C.-Y. Ed. American Chemical Society:.Washington DC, **1992**, pp. 72-86.
21. Vösgen, B. Herrmann, K. *Z. Lebensm. -Unters. -Forsch.* **1980**, *170*, pp. 204-207.
22. Nakatani, N.; Inatani, R.; Ohta, H.; Nishioka, A. *Environ. Health Perspect.* **1986**, *67*, pp. 135-142.
23. Farag, R.S.; Badei, A.Z.M.A.; Hewedi, F.M.; El-Baroty, G.S.A. *J. Am. Oil Chem. Soc.* **1989**, *66*, pp. 792-799.
24. Kramer, R.E. *J. Am. Oil Chem. Soc.* **1985**, *62*, pp. 111-113.
25. Deighton, N.; Glidewell, S.M.; Deans, S.G.; Goodman, B.A. *J. Sci. Food Agric.* **1993**, *63*, pp. 221-225.
26. Yoshikawa, T.; Naito, Y.; Tanigawa, T.; Kondo, M. *Arzneim. -Forsch. /Drug Res.* **1993**, *43*, pp. 363-366.

27. Chan, W.K.M.; Decker, E.A.; Lee, J.B.; Butterfield, D.A. *J. Agric. Food Chem.* **1994**, *42*, pp. 1407-1410.
28. Madsen, H.L.; Nielsen, B.R.; Bertelsen, G.; Skibsted, L.H. *Food Chemistry. In press.*
29. Nagababu, E.; Lakshmaiah, N. *Biochem. Pharmacol.* **1992**, *43*, pp. 2393-2400.
30. Tsujimoto, Y.; Hashizume, H.; Yamazaki, M. *Int. J. Biochem.* **1993**, *25*, pp. 491-494.
31. Aruoma, O.I.; Halliwell, B.; Aeschbach, R.; Löliger, J. *Xenobiotica* **1992**, *22*, pp. 257-268.
32. Frankel, E.N.; Huang, S.-W.; Aeschbach, R.; Prior, E. *J. Agric. Food Chem.* **1996**, *44*, pp. 131-135.
33. Frankel, E.N.; Huang, S.-W.; Kanner, J.; German, J.B. *J. Agric. Food Chem.* **1994**, *42*, pp. 1054-1059.
34. Svoboda, K.P.; Deans, S.G. *Flavour and Fragrance Journal* **1992**, *7*, pp. 81-87.
35. Chang, S.S.; Ostric-Matijasevic, B.; Hsieh, O.A.L.; Huang, C.-L. *J. Food Sci.* **1977**, 42, pp. 1102-1106.
36. Palitzsch, A.; Schulze, H.; Lotter, G.; Steichele, A. *Fleischwirtsch.* **1974**, *54*, pp. 63-68.
37. Djarmati, Z.; Jankov, R.M.; Schwirtlich, E.; Djulinac, B.; Djordjevic, A. *J. Am. Oil Chem. Soc.* **1991**, *68*, pp. 731-734.
38. Wada, S.; Fang, X. *Journal of Food Processing and Preservation* **1992**, *16*, pp. 263-274.
39. Fang, X.; Wada, S. *Food Research International* **1993**, *26*, pp. 405-411.
40. Ternes, W.; Schwarz, K. *Z. Lebensm. -Unters. -Forsch.* **1995**, *201*, pp. 548-550.
41. Palic, A.; Krizanec, D.; Dikanovic-Lucan, Z. *Fleischwirtsch.* **1993**, *73*, pp. 670-672.

Chapter 15

Antioxidative Effect and Kinetics Study of Capsanthin on the Chlorophyll-Sensitized Photooxidation of Soybean Oil and Selected Flavor Compounds

Chung-Wen Chen, Tung Ching Lee, and Chi-Tang Ho

Department of Food Science, Cook College, Rutgers, The State University of New Jersey, New Brunswick, NJ 08903–0231

The antioxidative effect of capsanthin and lutein on the chlorophyll-sensitized photooxidation of 2-ethylfuran, 2,4,5-trimethyloxazole, and 2,5-dimethyl-4-hydroxy-3(2H)-furanone (DMHF) was studied using a spectrophotometric method. Both capsanthin and lutein exhibited the antiphotooxidative effect on these flavor compounds. The antiphotooxidative activity of capsanthin was higher than that of lutein. The antioxidative effect of capsanthin, β-carotene, and lutein on the photooxidation of soybean oil containing chlorophyll was studied by the Rancimat method, in which the induction time and the anti-photooxidation index (API) were measured. The induction times of soybean oil which contained the carotenoids were longer than that of the control sample which contained no carotenoid. As the number of conjugated double bonds increased, the API of soybean oil increased. The capsanthin, which contains 11 conjugated double bonds, a conjugated keto group and a cyclopentane ring, had higher API than β-carotene, which contains 11 conjugated double bonds but neither a conjugated keto group nor a cyclopentane ring. The steady-state kinetics study of capsanthin on the chlorophyll-sensitized photooxidation of soybean oil was conducted by oxygen depletion method. The result showed that capsanthin quenched singlet oxygen only and its quenching rate constant was $5.746 \times 10^9 \ M^{-1}s^{-1}$ in methylene chloride.

Some food constituents are sensitive to photooxidation. Oil and fats are the most sensitive food constituents for photooxidation because they contain a large number of double bonds in their structures and probably due to the greater solubility of molecular oxygen in the lipid phase as compared to the aqueous phase (1,2). Flavor compounds especially five membered heterocyclic flavor compounds such as alkylfuran, alkylpyrrole and alkyloxazole compounds easily undergo photooxidation reaction (3).

The whole photosensitized oxidation reaction can be described as follows (*4*):

$$Sen \xrightarrow{hv} {}^1Sen^* \xrightarrow{ISC} {}^3Sen^* \longrightarrow {}^3O_2 \longrightarrow {}^1O_2 \xrightarrow{A} AO_2$$

A sensitizer, such as chlorophyll, can absorb light energy and becomes an excited singlet sensitizer ($^1Sen^*$), which is then rapidly converted to the excited triplet sensitizer ($^3Sen^*$) by an intersystem crossing (ISC) mechanism, a process of electron rearrangement. The $^3Sen^*$ transfers its energy to triplet oxygen ($^3O_2$) to generate the singlet oxygen ($^1O_2$) by a triplet-triplet annihilation reaction, a process which occurs via a collision of two excited $^3O_2$. Singlet oxygen is a very reactive form of oxygen and easily attacks the substrates (A) which have double bonds to generate oxidized products (AO_2).

The process of photosensitized oxidation can be minimized by quenchers, which can quench singlet oxygen and/or excited sensitizers. The antiphotooxidative effect of nickel chelates on the singlet oxygen oxidation of soybean oil have been reported by Lee and Min (*5*). Ascorbic acid and tocopherols are not only free radical scavengers but also good singlet oxygen quenchers (*6,7*). *tert*-Butyl hydroquinone (TBHQ) as well as other phenolic compounds can also act as singlet oxygen quenchers (*8-12*). Rosemary oleoresin and rosemariquinone, a rosemary extract, can inhibit light-induced oxidation (*13,14*). Among these quenchers, carotenoids appeared to be the most plausible candidates for quenching the singlet oxygen. The quenching mechanisms and kinetics of several carotenoids, including β-apo-8'-carotenal, β-carotene, canthaxanthin, lutein, zeaxanthin, lycopene, isozeaxanthin, and astaxanthin have been well studied (*15-17*).

Capsanthin (Figure 1) is the most abundant carotenoid in the paprika spice. The concentration of capsanthin is about 1590 mg per kg of dry matter and 41 ~ 54% of total carotenoids in paprika (*18,19*). The antiphotooxidative effect and kintics study of capsanthin on soybean oil has not been well studied. In addition, the study of the antiphotooxidative effect of carotenoids mostly focused on lipid or fatty acids. The antioxidative effect of carotenoids on the photooxidation of flavor compounds has not been well studied, either.

This paper reports the antioxidative effect of capsanthin on the chlorophyll-photosensitized oxidation of soybean oil and selected flavor compounds including 2-ethylfuran, 2,4,5-trimethyloxazole, and 2,5-dimethyl-4-hydroxy-3(2H)-furanone (DMHF), which are unstable when exposed to light in the presence of sensitizer (*3,20*). The quenching mechanism and kintics of capsanthin on the photosensitized oxidation of soybean oil are also reported. β-carotene and lutein were used as controls for comparing the antiphotooxidative activity with capsanthin.

EXPERIMENTAL

Materials. Soybean oil was purchased from a local supermarket and purified in our laboratory according to the method of Lee and Min (*15*). Chlorophyll was extracted and purified from spinach according to the method of Omata and Murata (*21*). ß-carotene was purchased from Sigma Chemical Co. (St. Louis, MO). Capsanthin,

Compound	Structure	No. of conjugated double bond
Capsanthin		11
ß-Carotene		11
Lutein		10

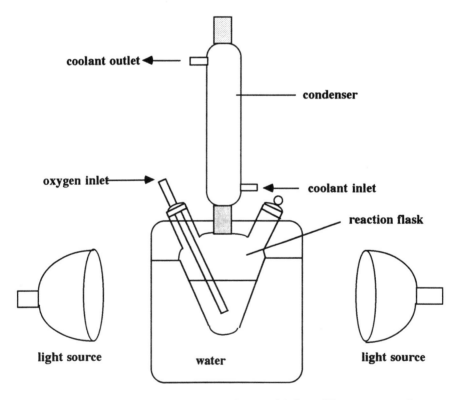

Figure 1. Structures of capsanthin, β-carotene, and lutein.

Figure 2. Apparatus for the photosensitized oxidation of flavor compounds.

which was extracted from paprika spice, and lutein were obtained from KALSEC Co. (Kalamazoo, MI). 2-Ethylfuran, 2,4,5-trimethyloxazole, and DMHF were purchased from Aldrich Chemical Co. (Milwaukee, WI). Absolute alcohol was obtained from Pharmco Products Co. (Brookfield, CT).

Evaluation of the antioxidative effect of carotenoids on the chlorophyll-sensitized photooxidation of 2-ethylfuran, 2,4,5-trimethyloxazole, and DMHF by spectrophotometric method. Mixtures of 1×10^{-3} M 2-ethylfuran, 2,4,5-trimethyloxazole, or DMHF in 100 ml absolute alcohol containing 10 ppm chlorophyll and either 1.0×10^{-5} or 2.5×10^{-5} M carotenoids were prepared in 250 ml of flask. The antiphotooxidative effect of carotenoids including capsanthin and lutein on these flavor compounds was carried out under light exposure at 20°C while bubbling with oxygen at 260 ml/min according to the method of Chen and Ho (3) as shown in Figure 2. The light intensity at the sample position was 67,000 lux. The effect of carotenoids was determined using a Hitachi U-3110 spectrophotometer (Danbury, CT) in which the percentage of substrate remaining was calculated by measuring the decrease of optimum wavelength absorbance of test compounds. The optimum wavelength absorbance of 2-ethylfuran, 2,4,5-trimethyloxazole, and DMHF was 215, 225, and 290nm, respectively.

Evaluation of the antioxidative effect of carotenoids on the chlorophyll-sensitized photooxidation of purified soybean oil by the Rancimat method. Samples of 100 or 200 ppm carotenoids in 2.5 g purified soybean oil containing 200 ppm chlorophyll were prepared in 2.0 cm i.d. x 15cm glass open cylinders. The cylinders were then placed in the light box, which was modified from the device as reported by Lee and Min (15). The light intensity at the sample level was 10,000 lux. The temperature of the light storage box was kept at 25°C. After 4 hrs storage in the light box, the cylinders were placed in a Model 679 Rancimat instrument (Metrohm, Switzerland). The oxidative stability of this soybean oil was measured by determining the induction time at 100°C. The air supply was maintained at 20 l/hr.

Quenching mechanism and kinetics study of capsanthin by oxygen depletion method. The quenching mechanism and kinetics of capsanthin were studied using the steady state kinetic method of Foote and Denny (22) and Lee and Min (16). Samples of 0.03, 0.06, 0.10 or 0.16 M purified soybean oil in methylene chloride containing 4 ppm chlorophyll and 0, 0.25×10^{-5}, 0.5×10^{-5}, 0.75×10^{-5} or 1×10^{-5} M capsanthin were prepared. Fifteen ml of prepared sample was transferred into 30ml serum bottles and sealed air-tight with rubber septa. Duplicate samples were placed in the light box for 2hrs at 25°C and the percentage of headspace oxygen was analyzed using a Hewlett-Packard 5890 II gas chromatograph (GC) equipped with a thermal conductivity detector (TCD) and a Hewlett-Packard 3396A integrator. The GC column was a two-layer concentric column (Alltech CTRI, Avondale, PA); the outer column was 6 ft x 1/4 in. o.d. packed with an activated molecular sieve; the inner column was 6 ft x 1/8 in. o.d. packed with a porous polymer mixture. The flow rate of helium carrier gas was at 20 ml/min. The temperature of the injection port, oven and detector were 50, 35 and 100°C, respectively.

RESULTS AND DICUSSION

Antiphotooxidative effect of carotenoids on 2-ethylfuran, 2,4,5-trimethyloxazole, and DMHF. The effect of 1.0×10^{-5} or 2.5×10^{-5} M capsanthin or lutein on the photooxidative stability of 2-ethylfuran, 2,4,5-trimethyloxazole, and DMHF in absolute alcohol containing chlorophyll was conducted under light exposure while bubbling with oxygen. The percentage of substrate remaining was determined using spectrophotometry and was used as an index of antiphotooxidative activity of carotenoids. The greater the percentage of substrate remaining, the higher the antiphotooxidative activity exhibited.

The antiphotooxidative effect of capsanthin and lutein on the 2-ethylfuran, 2,4,5-trimethyloxazole, and DMHF is shown in Figure 3. Both capsanthin and lutein showed the antiphotooxidative effect on these flavor compounds. The antiphotooxidative activity of carotenoids increased when the concentration increased from 1.0×10^{-5} to 2.5×10^{-5} M . Comparison of the same concentration of capsanthin and lutien, the antiphotooxidative activity of capsanthin was higher than that of lutein.

Antiphotooxidative effect of carotenoids on the induction time of soybean oil as measured by the Rancimat method. The effect of 100 ppm or 200 ppm capsanthin, β-carotene, or lutein on the photooxidative stability of soybean oil containing 200 ppm chlorophyll was measured by the Rancimat method after 4 hours light exposure at 25°C. Preliminary studies showed that the induction time of purified soybean oil containing 100 or 200 ppm carotenoids without chlorophyll was slightly lower than that of the blank sample which contained no carotenoids (data not shown). Therefore, there is no anti-autoxidative effect of carotenoids on the soybean oil. Table I shows that as the concentration of the carotenoids increased from 100 to 200 ppm, the induction time as well as the anti-photooxidation index (API) increased. The induction time of soybean oil containing the carotenoids was longer than that of the control sample which contained no carotenoid; however, it was still shorter than that of the blank sample which contained no carotenoid and no chlorophyll.

Table II shows that as the number of conjugated double bond increased from 10 to 11, the API of soybean oil increased at the concentrations of 100 and 200 ppm carotenoids. It has been reported that as the number of conjugated double bonds of carotenoids increased, the peroxide values of chlorophyll-sensitized photooxidation of soybean oil decreased significantly (16,17). Therefore, the singlet oxygen quenching ability of carotenoids was dependent on the number of conjugated double bonds of the carotenoids as reported by Lee and Min (16), Jung and Min (17), and Hirayama et al. (23).

Table II also shows that the soybean oil containing capsanthin had a higher API than the oil containing β-carotene although both of the carotenoids have the

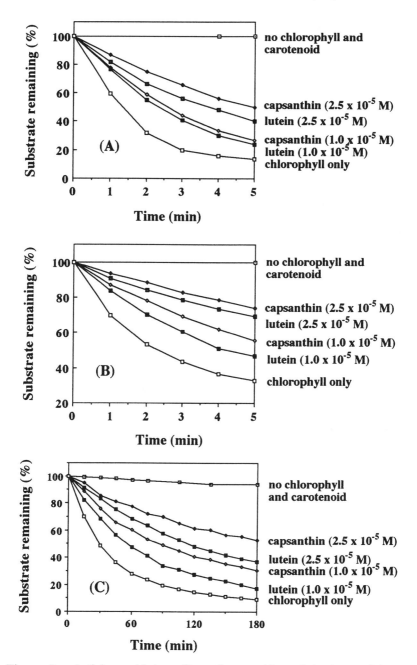

Figure 3. Antiphotooxidative effect of capsanthin and lutein on (A) 2-ethylfuran, (B) 2,4,5-trimethyloxazole, and (C) 2,5-dimethyl-5-hydroxy-3(2H)-furanone (DMHF) in absolute alcohol containing 10 ppm chlorophyll at 20 °C.

Table I. Effect of carotenoids on the induction time detected by the Rancimat method of soybean oil containing 200 ppm chlorophyll during light exposure for 4 hours.

Carotenoid (ppm)	Induction time (IT)[1] (hrs)	Anti-photooxidation index (API)[1]
Control[2]	7.96	-
Capsanthin (100)	8.94	0.303
Capsanthin (200)	9.64	0.518
Blank[3]	11.20	-
Control[2]	7.96	-
β-Carotene (100)	8.90	0.292
β-Carotene (200)	9.60	0.505
Blank[3]	11.20	-
Control[2]	7.96	-
Lutein (100)	8.74	0.241
Lutein (200)	9.39	0.440
Blank[3]	11.20	-

1. API = (carotenoid IT - control IT) / (blank IT - control IT)
   Each value is the means of duplicates
2. soybean oil + 200ppm chlorophyll
3. soybean oil only

Table II. Effect of the number of conjugated double bond of carotenoids on the anti-photooxidation index (API) of soybean oil during light exposure.

Carotenoid	Number of conjugated double boud	API
(100 ppm)		
Lutein	10	0.241
β-Carotene	11	0.292
Capsanthin	11	0.303
(200 ppm)		
Lutein	10	0.440
β-Carotene	11	0.505
Capsanthin	11	0.518

same number of conjugated double bonds. It has been reported that the singlet oxygen quenching ability of carotenoids was determined not only by the number of conjugated double bonds, but by the functional groups of the carotenoid (*23*). As reported by Hirayama et al (*23*), conjugated keto groups and the presence of a cyclopentane ring in the carotenoids stimulate quenching efficiency. Comparison of the structure of capsanthin with that of β-carotene reveals that there is one conjugated keto group and one cyclopentane ring in the structure of capsanthin but β-carotene contains neither of them (Figure 1). Therefore, the antiphotooxidative activity of capsanthin is higher than that of β-carotene, as shown in Table II. This result confirms Hirayama's prediction.

The Rancimat method has been used to measure the antioxidant activity of synthetic and natural antioxidants (*24-26*) and has correlated well with oil stability measured by the Active Oxygen Method (*27*) and peroxide value measurement (*28*). Our study showed that using the Rancimat method to study the antiphotooxidative effect of carotenoids on the soybean oil was in agreement with the results using the headspace oxygen depletion method (*16*) and the peroxide value method (*16,17*).

Quenching mechanism and kinetics of capsanthin. The following steady-state kinetic equation would be established if carotenoids reduced the chlorophyll-photosensitized singlet oxygen oxidation by singlet oxygen quenching (*4,22*).

$$\{-d[O_2]/dt\}^{-1} = \{d[AO_2]/dt\}^{-1}$$
$$= K^{-1}\{1 + (k_q[Q] + k_{OX-Q}[Q] + k_d)/k_r[A]\}$$

AO_2 : oxidized soybean oil
K : rate of singlet oxygen formation
k_q : reaction rate constant of physical singlet oxygen quenching by capsanthin
Q : capsanthin
k_{OX-Q} : reaction rate constant of chemical singlet oxygen quenching by carotenoid
k_d : decaying rate constant of singlet oxygen
k_r : reaction rate constant of soybean oil with singlet oxygen
A : soybean oil.

The intercepts and slopes of the plots of $(-d[O_2]/dt)^{-1}$ vs $[A]^{-1}$ at various concentrations of quencher (Q) are K^{-1} and $K^{-1}\{(k_d + k_q[Q] + k_{OX-Q}[Q]/k_r\}$, respectively. The plot of S_Q/S_O (slopes in the presence and absence of quencher) vs [Q] is a straight line, and the slope of the straight line is $(k_q + k_{OX-Q})/k_d$ (*4,29,30*). The rate constants (k_d) of singlet oxygen decay in different solvents have been reported (*31*). Therefore, the total singlet oxygen quenching rate constant ($k_q + k_{OX-Q}$) of quencher can be determined from the slope of the plot of S_Q/S_O vs [Q] [*4*].

The plots of $(-d[O_2]/dt)^{-1}$ vs [soybean oil]$^{-1}$ at different levels of capsanthin are shown in Figure 4. The intercepts were the same at different levels of the capsanthin, but the slopes of the plots increased as the concentration of capsanthin increased from 0 to 1×10^{-5} M. This result showed that capsanthin as well as other

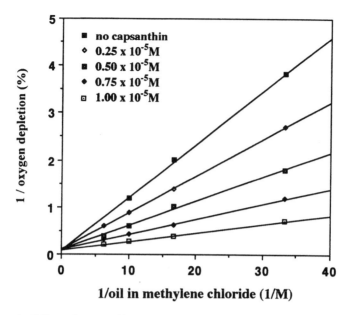

Figure 4. Effect of capsanthin on the headspace oxygen depletion of soybean oil in methylene chloride containing 4 ppm chlorophyll when exposed to light for 2 hours at 25 °C.

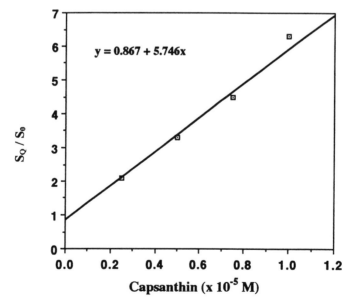

Figure 5. Plot of S_Q/S_0 vs [capsanthin].

carotenoids quenched the singlet oxygen but did not quench the excited triplet state of chlorophyll to reduce the photosensitized oxidation of soybean oil (*4,5,15,16*).

To measure the total singlet oxygen quenching rate constant ($k_q + k_{OX-Q}$) of capsanthin, the regression line of S_Q/S_O vs [capsanthin] was plotted as shown in Figure 5. S_Q and S_O are the slopes of the plot of $(-d[O_2]/dt)^{-1}$ vs [soybean oil]$^{-1}$ in the presence and absence of capsanthin, respcetively. As shown in Figure 5, the slope of the plot of S_Q/S_O vs [capsanthin] is 5.746 x 10^5 M^{-1}. Since the slope of the plot of S_Q/S_O vs [quencher] is ($k_q + k_{OX-Q}$)/k_d (*4,22,30*) and the k_d value of singlet oxygen in methylene chloride is 1.0 x 10^4 s^{-1} (*31*), the total singlet oxygen quenching rate constant ($k_q + k_{OX-Q}$) of capsanthin is 5.746 x 10^9 M^{-1}s^{-1} in methylene chloride.

CONCLUSION

It has been known that carotenoids are able to quench singlet oxygen to reduce the photosensitized oxidation reaction (*16,17*). Our study showed that capsanthin, a major carotenoid from paprika, as well as β-carotene and lutein exhibited the ability to quench singlet oxygen to reduce the photosensitized oxidation of both soybean oil and selected flavor compounds including 2-ethylfuran, 2,4,5-trimethyloxazole, and DMHF. These results suggest that capsanthin as well as other carotenoids may be applied to food system which contains food lipids or flavor compounds to minimize food photodeterioration.

REFERENCES

1. Rosenthal, I. in *Singlet O₂; Vol IV Polymers and Biomolecules,* Frimer, A.A., Ed.; CRC Press: Florida, **1985**, pp. 145-163.
2. Samuel, D.; Steckel, F. in *Molecular Oxygen in Biology,* Hayaishi, O., Ed.; Elsevier: New York, **1974**, 1.
3. Chen, C.-W.; Ho, C.-T. *J. Agric. Food Chem.* in press.
4. Foote, C.S. in *Singlet Oxygen,* Wasserman, H.H. and Murray, R.W., Eds.; Academic Press: New York, **1979**, pp. 139-171.
5. Lee, S-.H.; Min, D.B. *J. Agric. Food Chem.* **1991**, *39*, 642-646.
6. Jung, M.Y.; Choe, E.; Min, D.M. *J. Food Sci.* **1991**, *56*, 807.
7. Chou, P-.T.; Khan, A.U. *Biochem. Biophys. Res. Comm.* **1983**, *115*, 932-937.
8. Sherwin, E.R. in *Food Additives.* Branen, P.M.; Davidson and Salminen,S., Eds.; Marcel Dekker, Inc; New York. **1990**, p. 139.
9. Clough, R.L.; Yee, B.G.; Foote, C.S. *J. Amer. Chem. Soc.* **1979**, *101*, 83.
10. Matsuura, T.; Yoshimura, N.; Nishinaga, A.; Saito, I. *Tetra. Lett.* **1969**, *21*, 1669-1671.
11. Thomas, M.J.; Foote, C.S. *Photochem. Photobiol.* **1978**, *27*, 683-693.
12. Tournaire, C.; Croux, S.; Maurette, M-.T. *J. Photochem. Photobiol.* **1993**, *19*, 205-215.
13. Hall III, C.; Cuppett, S. *J. Amer. Oil Chem. Soc.* **1993**, *70*, 477-482.

14. Hall III, C.; Cuppett, S.; Wheeler, D.; Fu, X. *J. Amer. Oil Chem. Soc.* **1994**, *71*, 533-535.
15. Lee, E.C.; Min, D.B. *J. Food Sci.* **1988**, *53*, 1894-1895.
16. Lee, S-.W.; Min, D.B. *J. Agric. Food Chem.* **1990**, *38*, 1630-1634.
17. Jung, M.Y.; Min, D.B. *J. Amer. Oil Chem.* Soc. **1991**, *68*, 653-658.
18. Minguez-Mosquera, M.I.; Hornero-Méndez, D. *J. Agric. Food Chem.***1994**, *42*, 1555-1560.
19. Biacs, P.A.; Daood, H.G.; Huszka, T.T.; Biacs, P.K. *J. Agric. Food Chem.* **1993**, *41*, 1864-1867.
20. Chen, C.-.W.; Shu, C.-K.; Ho, C.-T. *J. Agric. Food Chem.* in press.
21. Omata, T.; Murata, N. *Photochem. Photobiol.* **1980**, *31*, 183-185.
22. Foote, C.S.; Denny, R.W. *J. Amer. Chem. Soc.* **1968**, 90, 6233-6235.
23. Hirayama, O.; Nakamura,K.; Hamada, S.; Kobayasi, Y. *Lipids* **1994**, *29*, 149-150.
24. Chen, C.-W.; Ho, C.-T. *J. Food Lipids* **1995**, *2*, 35-46.
25. Zhang, K.Q.; Bao, Y.; Wu, P.; Rosen, R.T.; Ho, C.-T. *J. Agric. Food Chem.* **1990**, *38*, 1194-1197.
26. Ho, C-.T.; Chen, Q.; Shi, H.; Zhang, K.Q.; Rosen, R.T. *Preventive Med.* **1992**, *21*, 520-525.
27. Laubli, M.W.; Bruttel, P.A. *J. Amer. Oil Chem. Soc.* **1986**, *63*, 792-795.
28. Gordon, M.H.; Mursi, E. *J. Amer. Oil Chem. Soc.* **1994**, *71*, 649-651.
29. Foote, C.S.; Ching, Ta-Yen.; Geller, G.G. *Photochem. Photobiol.* **1974**, 20, 511-513.
30. Yamauchi, R.; Matsushita, S. *Agric. Biol. Chem.* **1977**, 41, 1425-1430.
31. Hurst, J.R.; McDonald, J.D.; Schuster, G.B. *J. Amer. Chem. Soc.* **1982**, 104, 2065-2067.

Chapter 16

Curcumin: An Ingredient that Reduces Platelet Aggregation and Hyperlipidemia, and Enhances Antioxidant and Immune Functions

Yaguang Liu

L. Y. Research Corporation, 67–08 106A Street, Flushing, NY 11365

The effects of curcumin on blood platelet aggregation, hyperlipidemia, lipid peroxidation and immune function were determined. Curcumin exhibited strong inhibition of blood platelet aggregation in humans and lipid peroxidation induced by adenosine diphosphate (ADP) of rats. The marked rise in serum cholesterol levels in rats following 4 weeks of cholesterol feeding was significantly reduced by curcumin. Additionally, curcumin showed that it increased immune function, including macrophage rate and lymphoblastoid transformation in immunosuppression rats induced by cyclophosphane (CY). The above pharmaceutical effects of curcumin with addition of flavone of *Matricaria L.* were stronger than curcumin alone. It can be seen that curcumin effectively prevents cardiovascular disease, increases immune function and exhibits anti-oxidant activity.

Curcumin is isolated from the rhizomes of the plant, *Curcuma longa L.* which is a medicinal plant widely used in China, India and Southeast Asia. The rhizomes of the plant has been used in China as a reduction in serum hyperlipidemia and stomachic drug *(1)*.

Curcumin has demonstrated anti-inflammatory and anti-oxidant properties and it is essentially non-toxic *(2-6)*. Recent studies have indicated that a group of four ingredients, which include curcumin, scoparone, ferulic acid and flavone of *Matricaria L.* has demonstrated marked hyperlipidemia and platelet-aggregation reducing properties *(7)*.

This paper describes the effects of curcumin and flavone of *M. chamomilla L.* on platelet-aggregation, hyperlipidemia, anti-oxidant and immune functions.

Materials

Curcumin was extracted from *C. longa L.* according to the following steps.

Rhizome of *C. longa L.* or *C. aromatica salisb* or *C. zedoaria Rosco* was dried and powered. One kilogram of the powder was soaked in 5 liters of 95% ethanol, for 12 hours at room temperature. The resulting ethanol extract was filtered. The ethanol was then recovered under reduced pressure distillation. A residue was dissolved in 200 ml of 1 N NaOH. The resulting solution was adjusted to pH 7 with 1 N HCl and a light yellow precipitate was formed. The precipitate was dissolved in 300 ml of 95% ethanol and the last procedure was repeated to form a light yellow precipitate again. The light yellow precipitate was washed with acetone and ether successively and then dried under vacuum and was found to have a melting point at $183\,^0C$.

Flavone was extracted from *M. chamomilla L.* according to the following steps.

Plants of *M. chamomilla L.* were dried and powdered. One kg of the powder was extracted in 2 liters of 95% ethanol for about 24 hours at room temperature. The solution was filtered and the filtrate saved. Two liters of 95% ethanol were added to the residue and refluxed in a water bath for 6 hours. The refluxed ethanol was cooled and filtered and the filtrate combined with the extract filtrate. Ethanol was then recovered by reduced pressure distillation and the residue was saved. One thousand ml of acetic ether was added to the residue and refluxed in a water bath for 6 hours. The refluxing procedure was repeated. Acetic ether was then concentrated under reduced pressure distillation. Crystals were formed which were then washed with water. The final crystals were dried under vacuum and were found to have a melting point of about 250^0C *(7)*.

C. longa L. and *M. chamomilla L.* are recognized by Food and Drug administration (FDA) as safe for human consumption.

Methods

Blood platelet Aggregation. The blood platelet aggregation test was performed according to the method of Born et al. *(8)*. Blood was collected from veins of humans using a needle attached to a plastic disposal syringe. The blood was immediately transferred to a siliconized glass tube containing 0.1 volume of 3.13% sodium citrate. Platelet-rich plasma (PRP) was obtained by centrifugation of the whole blood at 1000 rpm for 10 min at room temperature. Platelet-poor plasma (PPP) was prepared by centrifugation of the remaining blood at 3000 rpm for 10 min. Platelet aggregation was performed at 37^0C. Human platelet studies were carried out with a constant platelet number (3×10^8 /ml). With regard to determination of platelet aggregation, maximum aggregation was induced by 0.2 μM (ADP). 0.4 ml PRP was added into each tube. Thirty six tubes were divided into three groups. Fifty μl of saline was added to each tube of control group, 50 μl of curcumin (0.5 mg/ml) was added to each tube of the crucumin (treatment 1) group, and 50 μl of curcumin (0.45 mg/ml) + 5 μl flavone of *M. chamomilla L.* (0.05 mg/ml) were added to each tube of curcumin plus flavone of *M. chamomilla L.* (treatment 2) group. After incubation for 3 minutes at 37^0 C., 50 μl of 2 M ADP was added to each tube. A five minute's aggregation curve for each tube was plotted.

Percent inhibition of aggregation by PHP was calculated by:

$$\% \text{ inhibition of aggregation} = \frac{\% \text{ aggregation in control} - \% \text{ aggregation with treatment}}{\% \text{ aggregation in control}} \times 100 \%$$

Determination of Lipid Peroxidation in Mitochondria and Microsomes. Rats were killed by decapitation after fasting for 24 hours and their liver tissue was quickly removed. Microsomal and mitochondria fractions were isolated from the liver tissue by the method of Oda *(9)*. The liver tissues were cut into small slices in 0.25 M sucrose containing 3 mM Tris-HCl and 0.1 mM EDTA (pH 7.4) at 4^0C, and then small slices of liver tissues were homogenized with nine times (the amount by weight) 0.25 M sucrose solution containing 3 mM Tris-HCl and 0.1 mM EDTA (pH 7.4) at 4^0 C. The homogenate solutions were adjusted to pH 7.4 by the addition of 0.1 N KCl, and then the homogenate was centrifuged at 50 x g for 10 minutes at 4^0 C to remove the nuclear fractions and red blood cells. The supernatant phase was centrifuged at 700 x g for 10 minutes at 4^0C, and then further centrifuged at 5,000 x g for 45 minutes at 4^0C to give mitochondrial fractions. The isolated mitochondrial fractions were washed twice with Krebs-Ringer phosphate buffer (pH 7.4). The supernatant phase centrifuged at 5,000 x g for 45 minutes was further centrifuged at 24,000 x g for 10 minutes at 4^0 C. The precipitation was removed and the supernatant phase was again centrifuged at 54,000 x g for 60 minutes at 4^0 C to obtain microsomal fractions. The isolated microsomal fractions were washed twice with Krebs-Ringer phosphate buffer (pH 7.4). Mitochondria and microsomes were suspended in Krebs-Ringer phosphate buffer (pH 7.4). The final protein concentrations of the mitochondrial and microsomal suspensions were adjusted to 10 mg protein/ml.

A mixture of mitochondrial suspension (0.5 ml, 10 mg protein/ml), Kerbs-Ringer phosphate buffer containing 40 mM ADP, 12 mM ascorbic acid and the indicated amount of curcumin and flavone of *M. chamomilla L.* were incubated at 37^0 C for 15 minutes in a final volume of 1 ml. In the treatment 1 group, buffer contained 5×10^{-4} M curcumin. In the treatment 2 group, buffer contained 0.45×10^{-4} M curcumin and 0.05×10^{-4} M flavone. Lipid peroxides of mitochondria were determined by the method of Ohkawa et al. *(10)*.

Reducing Hyperlipidemia Test. Young male Wistar rats with an average weight of 130 g were fed normal laboratory diet and were housed individually in cages. Room temperature was controlled at 25 ± 1^0 C with 60% relative humidity. Lighting was controlled with cycles of 12 hours light followed by 12 hours of dark.

Forty male rats were separated into four groups. The normal group was given a regular laboratory diet. Control and treatment groups 1 and 2 were given a regular diet supplemented with 1% cholesterol and 0.2% cholic acid. A restricted diet (8g) was prescribed every morning. Curcumin (dose 100 mg/kg/day) was dissolved in distilled water (2 ml/rat) and was orally administered to rats of treatment 1 group. Curcumin (dose 90 mg/kg/day) plus flavone (dose 10 mg/kg/day) were dissolved in distilled water (2 ml/rat) and was orally administered to rats of treatment 2 group. After 4 weeks of feeding, the rats were starved 18 hours, then sacrificed. Blood was collected from the jugular vein. The blood and liver were removed immediately and portions of the tissue were examined to determine cholesterol, triglyceride and total lipid.

Lipid Content. Total lipid contents of the liver and serum were determined by the methods of Sperry and Brand *(11)*. Total cholesterols were determined by the methods of Carr and Drekter *(12)*, free cholesterol by the method of Sperry *(13)*, and triglycerides by the method of Van handel-Zilversmit *(14)*.

Immune Function. Male mice weighting 18-20 g were used in the experiments and were divided into treatment 1, 2 and control groups. For treatment 1 group, the dosage of curcumin was 5.5 mg/kg injected intraperitoneally. For treatment 2 group, the dosage of curcumin was 5.0 mg/kg and 0.5 mg/kg flavone. The mice of control group were injected with the same volume of normal saline. These injections were repeated daily for 7 days. On the last day, both treated and control groups were injected intraperitoneally with CY. The dosage of CY was 4.5 mg/kg. On the seventh day, the animals were killed by decapitation. The blood was removed immediately and was examined to determine immune function.

Lymphoblastoid Transformation Test. Lymphoblastoid transformation test was performed using the method of Hogan *(15)*. Cells (1 x 10^6/ml) were maintained in Eagles minimum essential medium containing 10% fetal bovine serum (FBS) and 10 µg/ml polysaccahride-free purified phytohemagglutinin (PHA-P). The cells were incubated at 37^0C/5% CO_2 /95% humidity and the medium changed every day.

Macrophage. The macrophage cells were cultured in PPMI-1640 medium with 10% FBS. After overnight incubation at 37^0 C in an atmosphare of 5% CO_2 and 95% air, non-adherent cells were washed away. The adherent cells were further incubated with 4 ml of the medium for 24 hours at 37^0 C. Antigens were added to macrophage monolayers. After 6 hours incubation at 37^0 C, the monolayers were washed four times to remove antigens and immediately used for experiments. Macrophage cytotoxicity was measured as previously described *(16)*. Results are expressed as a specific percentage of ^{53}Cr release (percentage cytotoxicity) as calculated by the following formula: percentage specific cytotoxicity = 100 x (experimental c.p.m. - spontaneous c.p.m.)/(total c.p.m. -spontaneous c.p.m.).

Results and Discussion

The significant inhibitory effect of curcumin on platelet-aggregation is summarized in Table I below. Curcumin with flavone of *M. chamomilla L.* appeared to be more effective than curcumin alone. Platelet aggregation and subsequent adhesion under the influence of local flow irregularities are primary and injury of the intimal tissue is secondary *(17)*. Organ damage and dysfunction may be a consequence of platelet aggregation and embolization of platelet thrombi in micro-circulation. Aggregation of platelets might cause damage resulting in obstruction of vessels. Intermittent infusions of adenosine diphosphate into the coronary circulation of pigs caused circulatory collapse, electrocardiographic evidence of myocardial ischemia, ventricular dysrhythmias and formation of platelet aggregates in the microcirculation *(18, 19)*.

Table I. Effects of Curcumin on Aggregation of Platelet

	Rate of aggregation of platelet
Control	67.6 ± 5.0
Treatment 1	$26.8 \pm 0.8*$
Treatment 2	$24.7 \pm 0.7*$

Each value represents the mean of 10 animals ± S.E.
* $P<0.01$, significantly different from the normal group.
Treatment 1: Curcumin only.
Treatment 2: Curcumin plus flavone of *Matricaria chamomilla L.*

As shown in Table II, the blood cholesterol level in rats fed 1% cholesterol was elevated 284% and supplementation of curcumin and curcumin plus flavone of *M. chamomilla L.* to the cholesterol diet depressed this elevation significantly. Phospholipid levels showed no significant change. The addition of flavone of *M chamomilla L.* appeared to be more effective than the curcumin-only group.

Table II. Effect of Curcumin on Cholesterol Levels in Blood

Diet	Cholesterol (mg/dl)	Phospholipid (mg/dl)	Cholesterol/Phospholipid (%)
Normal	125 ± 6	201 ± 5	62
Cholesterol	480 ± 40	230 ± 9	209
Treatment 1	$210 \pm 10*$	$208 \pm 6*$	101
Treatment 2	$176 \pm 8*$	$195 \pm 5*$	90

As shown in Table III, total lipid levels, free and total cholesterol, and triglyceride levels of the liver decreased significantly when curcumin and flavone of *M. chamomilla L.* were added to the hypercholesterolemic diet but free cholesterol in the liver appeared no significant change.

Table III. Effect of Curcumin on Lipid Level

Group	Per 100 g liver tissue			
	Total lipid (g)	Total cholesterol (mg)	Free cholesterol (mg)	Triglyceride (mg)
Normal	7.5 ± 1.0	516 ± 90	231 ± 21	397 ± 25
Cholesterol	17.3 ± 2.1	2472 ± 151	327 ± 26	520 ± 27**
Treatment 1	9.0 ± 0.8*	1101 ± 92*	291 ± 27*	451 ± 25**
Treatment 2	8.1 ± 0.8*	921 ± 80*	280 ± 24*	411±23*

Each value represents the mean of 10 samples ± S.E.
*$P < 0.05$, significantly different from the normal group; ** $P < 0.01$

Recent clinical studies have shown that the plant *C. longa L.* significantly reduces blood lipid *(1)*. As shown in Table III, curcumin can also significantly reduce blood and liver cholesterol levels. Therefore, these studies suggest that curcumin may be the active ingredient responsible for the lipid-lowering effects of the plant *C. longa L.*

Table IV lists the lipid peroxidation in rat liver mitochondria. Treatment 1 and 2 significantly reduced the level of peroxidation. Peroxidation of lipids in the liver has been shown to result in many kinds of toxicity such as of membrane function and damage to membrane bound enzymes *(20)*.

Table IV. Effects of Curcumin on ADP plus Ascorbic Acid-induced Lipid Peroxidation in Rat Liver Mitochndria

Group	Inhibition
Normal	-----
Treatment 1	55.4% ± 4.5*
Treatment 2	69.1% ± 5.0*

Each value represents the mean of 10 samples ± S.E.
* $P<0.01$, significantly different from the normal group.

CY is a very strong immune inhibitor. It inhibited phagocytic rate to 77% and lymphoblastoid to 48% in rats. As shown in Table V and VI, curcumin and curcumin with flavone of *M. chamomilla L.* can significantly increase immune function which has been markedly decreased by CY.

Table V. Effects of Curcumin on Macrophages

Group	Phagocytic rate ± SD (%)
Control	35.10 ± 2.01
CY	8.00 ± 0.36
CY + Treatment 1	21.00 ± 0.20*
CY + Treatment 2	23.20 ± 0.20*

Table VI. Effect of Curcumin on Lymphoblastoid Transformation

Group	CPM ± SD
Control	1340 ± 51
CY	697 ± 38
CY + Treatment 1	905 ± 70*
CY + Treatment 2	985 ± 75*

Each value represents the mean of 10 samples ± S.E.
*P<0.01, significantly different from cyclophosphamide group.

On the basis of the above results, we conclude that curcumin may protect against cardiovascular disease, increase immune function and has anti-oxidant activity.

Literature Cited

1. Xue C.S. Pharmocology and application of Chinese Medicine; Wang Y. Ed.; People Health Published House: Beijing,1983; pp 849-850.
2. Sharma, O.P. Bioc. Pharm. 1976, 25, 1811-1812.
3. Toda, S.; Miyase, T.; Arichi, H.; Tanizawa, H.; Takino, Y. Chem. Pharm. Bull. 1985, 33, 1725-1728.
4. Srimal R. C.; and Dhawan, B.N. J. Phar. Pharmacol. 1973, 25, 447-452.
5. Rao, T. S.; Basu, N.; Siddiqui, H.H. Indian J. Med. Res. 1982, 75, 574-578.
6. Mukhopadhyay, A.; Basu, N.; Gujral, P.K. Agents and Actions 1982, 12, 508-515.
7. Yaguang Liu. U.S. patent #4,842,859, 1989, pp 2-3.
8. Born, F.V.R; Cross, M.J. J. Physiiol. 1963, 168,178.
9. Ohkawa, H; Ohishi, N; Yagi, K. Biochem. 1979, 95, 351.
10. Oda, T; Seki,S. J. Elec. Micro. 1965, 12, 210.
11. Sperry, W.M.; Brand, F.C. J. Biol Chem. 1955, 213, 69.
12. Carr, J.J.; Drekter, I. J. Clin. Chem. 1956, 2, 353.
13. Sperry, W.M.; Webb, M. J. Biol Chem. 1959, 187, 97.
14. Vanhandel, E.; Zilversmit, D.B. J. Lab Clin. Med. 1967, 50, 152.
15. Hogan, M.M.; Vogel, S.N. J. Immuno. 1987, 139, 3697-3703.
16. Schwartz, R.H.L., Jackson L.; Paul, W.E. J. Immunol. 1975, 115, 1330.
17. Jorgensen, L.; Haeren, J.W.; Moe, N. Throm. Diat. Haem. 1973, 29, 470.
18. Jorgensen, L.; Rowosll, H.G.; Hovig, T. Lab Invest. 1967, 17, 616.
19. Moore, S.; Merserreau, W.A. Art Pathol. 1968, 85, 623.
20. Rice-Evans, C.; Hochstein, P. Biochem. Biophys. Res. Comm. 1981, 100,1537.

Chapter 17

Antioxidant Activity of Lavandin (*Lavandula x intermedia*) Cell Cultures in Relation to Their Rosmarinic Acid Content

T. López-Arnaldos[1], J. M. Zapata[2], A. A. Calderón[1],
and A. Ros Barceló[1]

[1]Department of Plant Biology, University of Murcia,
E–30100 Murcia, Spain
[2]Department of Plant Biology, University of Alcalá de Henares,
E–28871 Alcalá de Henares, Spain

Antioxidants are one of the principal ingredients that protect food quality by preventing oxidative deterioration of biomolecules. There is a growing interest in the application of antioxidants from plant material, herbs and spices. Spices provide one of the most promising sources of antioxidant compounds. Extracts from the Lamiaceae and Boraginaceae families have been reported to be effective antioxidants. The antioxidant activity is primarily attributed to phenolic compounds that function as free radical terminators and, in some cases, also as metal chelators. The present study was undertaken to assess the antioxidant activity of extracts from lavandin (*Lavandula x intermedia*) cell cultures. We have compared the superoxide anion scavenger activity of rosmarinic acid (main constituent of lavandin extracts) with that displayed by other structurally-related compounds. Finally, the iron-chelating properties of rosmarinic acid are reported.

The food industry routinely uses synthetic antioxidants such as BHA, BHT and TBHQ to protect products from oxidative deterioration. Although these antioxidants are effective, highly stable and inexpensive, there is concern about potential adverse effects from synthetic antioxidants (*1*). Toxicological data support the safety of the synthetic antioxidants when used at normal use levels. Public perception that synthetic compounds are dangerous is driving the industry to search for "natural" antioxidants.

Traditionally spices have been used to improve flavor and stability of food. One mode of preservation by plant extracts is the antioxidant activity which protects foods against off flavor related to lipid oxidation. The antioxidative activity found in many plants is due to the presence of compounds of a phenolic nature (*1*).

Phenolic antioxidants function as free radical terminators and in some cases as metal chelators. Phenolic acids particularly have been identified as potent antioxidants. Caffeic acid and its esters are good examples of phenolic antioxidants (2).

Rosmarinic acid (α-O-caffeoyl-3,4-dihydroxyphenyllactic acid) (Figure 1) is an example of a caffeic acid ester occurring in plants. Rosmarinic acid is mainly found in species of the Boraginaceae and Lamiaceae families, but can also be detected in other families (i.e. Apiaceae), ferns and hornworts (Table I) (3,4). This suggests that the ability to synthesize this caffeoyl ester may actually be widespread as evidenced by rosmarinic acid accumulation in a range of species (Table I) (5).

Current research on rosmarinic acid centers on its physiological and pharmacological activities, and it is regarded as a potential pharmaceutical plant product (5). So, rosmarinic acid has been shown to suppress the complement-dependent components of endotoxin shocks in rabbits and the oxidized compound displays antithyrotropic activity on human thyroid membrane preparations (5). Furthermore, there is increasing concern about the suitability of rosmarinic acid as a food additive.

One approach to recovery of useful plant metabolites such as rosmarinic acid is to extract plants as they are currently produced. Unfortunately the agronomic conditions for production of such plants are extremely variable. Unreliable cultivation, harvesting, shipping techniques, infestation with pests and natural status of the plant can all impart the lost of the product produced by the plant (6). Changes in any of these conditions can adversely influence the production of compounds such as rosmarinic acid. The application of modern culturing of plant cells offers an alternative means of production. Conditions of culturing can be optimized such that a cell culture would produce rosmarinic acid on a continuous basis at high yield and in a form where it can be easily extracted (7).

Occurrence of Rosmarinic Acid in Lavandin Cell Cultures

Cultures have been successfully established from a number of plant species which accumulate rosmarinic acid in their tissues. However, only a few of these cultures have demonstrated the ability to continuously produce the ester *in vitro* (5). In 1994, López-Arnaldos et al. (8) established cell cultures derived from leaves and spikes of lavandin, a sterile hybrid between *Lavandula angustifolia* and *Lavandula intermedia*. Microscopic analyses of the cells forming the callus showed the presence of blue fluorescence in the vacuoles when the preparations were observed under UV light. This fluorescence is indicative of the presence of rosmarinic acid in the vacuoles. This result is in agreement with those of Häusler *et al.* (9), who have recently demonstrated the accumulation of rosmarinic acid in vacuoles of suspension-cultured cells of *Coleus* by protoplast and vacuole isolation (9).

To corroborate the presence of rosmarinic acid in lavandin cells cultured *in vitro*, 50 % (p/v) methanolic extracts were obtained from the cells. In Figure 2, the UV spectrum of non-purified methanolic extracts is practically superimposible on that of pure rosmarinic acid. This suggests that rosmarinic acid is the main phenolic present in the methanolic extracts of the lavandin cells. Evidence in support of this suggestion arises from the HPLC analyses of non-purified methanolic extracts obtained from lavandin cell cultures (Figure 3).

Preparative low pressure chromatography on Sephadex LH-20 (⌀ 26 x 420 mm; methanol as eluent) of methanolic extracts of the cells and subsequent analysis of the

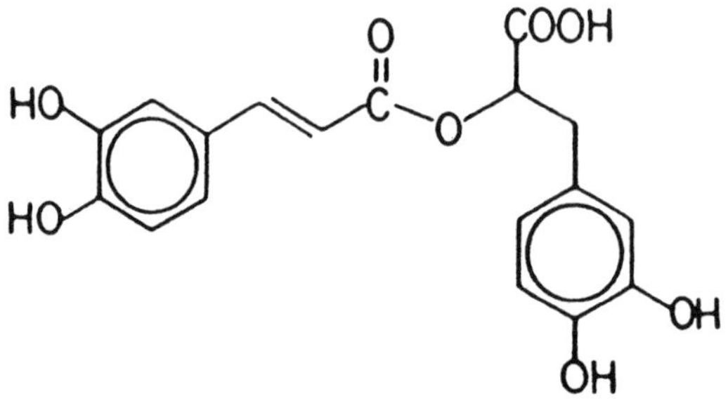

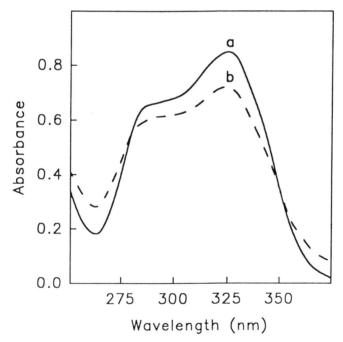

Figure 1. Structure of rosmarinic acid.

Figure 2. UV spectra of pure rosmarinic acid (a) and of non-purified methanolic extracts obtained from lavandin cell cultures (b).

Table I. Occurence of rosmarinic acid in some plant families

Plant	Organ	[RA] (%, w/w)
Apiaceae		
Eryngium campestre	leaf	0.8
Sanicula europaea	"	3.0
Boraginaceae		
Anchusa officinalis	flower	0.4
Borago officinalis	"	0.7
Lithospermum officinale	"	0.3
Pulmonaria officinalis	leaf	0.8
Lamiaceae		
Calamintha sylvatica s.l.	"	0.4
Lavandula angustifolia s.l.	"	1.2
Lavandula x burnatii	"	0.5
Lavandula latifolia	"	0.7
Lycopus europaeus	"	3.7
Melissa officinalis s.l.	"	4.7
Mentha x piperita	"	2.8
Mentha pulegium	"	2.7
Mentha spicata	"	3.9
Monarda didyma	flower	1.8
Nepeta cataria	"	0.2
Ocimum basilicum	"	1.9
Origanum vulgare	"	5.0
Orthosiphon aristatus	leaf	0.8
Prunella vulgaris	flower	6.1
Rosmarinus officinalis	"	2.5
Salvia lavandulifolia s.l.	leaf	0.6
Salvia officinalis	"	3.3
Satureja montana s.l.	flower	1.2
Thymus vulgaris	"	2.6
Zosteraceae		
Zoostera marina	root	1.9

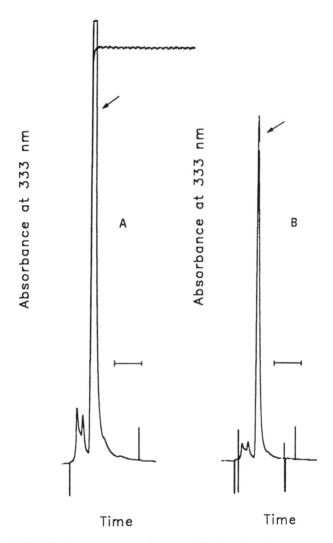

Figure 3. HPLC chromatograms of non-purified methanolic extracts obtained from cell cultures derived from leaves (A) and spikes (B) of lavandin (Bar = 4 minutes).

resultant fractions by TLC (silica gel; acetic acid:methanol:dichloromethane, 4:15:35 v/v/v) and UV-Visible spectroscopy revealed that, under the assay conditions, no phenolic other than rosmarinic acid could be detected after examination of TLC plates under UV light (λ=365 nm) and by spraying them with a solution of ferric chloride (2 %, p/v) (data not shown). Analyses of the plates showed that only a few fractions were enriched in rosmarinic acid, as judging by comparison of the R_f of the spots with that of pure rosmarinic acid. Furthermore, the UV spectra of these fractions were identical to that showed by pure rosmarinic acid (data not shown).

The determination by HPLC of rosmarinic acid in methanolic extracts of lavandin cell cultures revealed a content of about 3 micromoles per gram of fresh weight (*10*). This means a concentration of about 2-3 % in weight as regards the dry weight. Similar concentrations have been reported working with other plant materials (Table I) (*3*).

Although rosmarinic acid has been previously detected and quantified in field-grown plant material from the genus *Lavandula* (*3*), López-Arnaldos *et al.* (*8*) were the first to demonstrate production of rosmarinic acid in cell cultures derived from plants belonging to this genus. Further studies are necessary in order to optimize production and evaluate the potential for large-scale production of rosmarinic acid in lavandin cell cultures.

Superoxide anion-scavenging activity of extracts obtained from lavandin cell cultures

The presence of rosmarinic acid in methanolic extracts of lavandin cell cultures suggests that these extracts may offer potential as antioxidants. Despite the fact that lavandin belongs to the Lamiaceae family, there is little information about its antioxidant properties (*11*).

In order to determine antioxidative activity of compounds or extracts obtained from plants, several methods have been developed (*12*). Comparison of the results obtained by different techniques is complicated because antioxidative activity depends on several factors including the substrate used in the evaluation. However, for screening purposes, a model system is much less time-consuming than traditional storage studies (*1*).

In our study, the antioxidative activity of methanolic extracts from lavandin cell cultures was monitored by following the ability of these extracts to scavenge superoxide anions generated in a phenazine methosulphate (PMS)-NADH system. Figure 4 shows the absorbance at 270-280 nm (indicative of total phenolics) and at 325 nm (indicative of rosmarinic acid) as well as the inhibition of the superoxide dismutase-sensitive nitroblue tetrazolium chloride (NBT) reduction in fractions obtained after chromatography of lavandin methanolic extracts on Sephadex LH-20. From this figure, it can be observed that maximum superoxide anion scavenging ability is associated with those fractions containing rosmarinic acid. This fact suggests that rosmarinic acid is the main compound responsible for the superoxide anion-scavenging activity of methanolic extracts obtained from lavandin cell cultures.

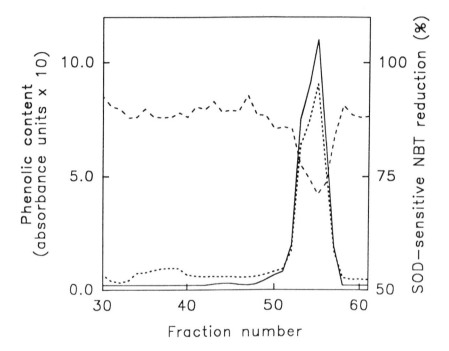

Figure 4. Total phenolic content, expressed as absorbance at 270-280 nm (···),
rosmarinic acid content, expressed as absorbance at 325 nm (—), and
superoxide dismutase-sensitive NBT reduction (---) of fractions obtained after
chromatographic separation on Sephadex LH-20 of methanolic extracts from
lavandin cell cultures.

Superoxide anion-scavenging activity of rosmarinic acid and other related compounds

To better understand how rosmarinic acid acts as a superoxide anion scavenger, the ability of other structurally-related phenolics to inhibit the NBT reduction in a PMS-NADH system was determined for comparison. The compounds tested were: (A) Protocatechuic acid (a dihydroxybenzoic acid), (B) Caffeic acid (a dihydroxycinnamic acid), (C) Chlorogenic acid (an ester of caffeic acid and quinic acid), and (D) Rosmarinic acid (which can be considered as a dimer of caffeic acid) (Figure 5). IC_{50} values for the scavenging of superoxide anions by phenolic compounds were calculated from the inhibition's percentage- concentration regression lines of the titrations represented in Figure 5.

Table II contains the IC_{50} values obtained from data in Figure 5. As can be seen in this Table, of all the phenolics tested, rosmarinic acid was the most efficient compound in deactivating the superoxide anion, inhibiting the superoxide dismutase-sensitive NBT reduction by 50 % with a concentration of 33 μM.

The antioxidative potency of a compound is related to its structure, in particular to electron delocalization of the aromatic nucleus and is also influenced by the number of hydroxyl groups (13). In this way, the higher antioxidative activity of rosmarinic acid with respect to that of the other phenolics tested could be related with the presence of two catechol groups on its molecule. However, since the IC_{50} value for rosmarinic acid is less than half of that for caffeic acid (Table II), other mechanisms for superoxide anion deactivation may be involved.

When other assays are used in order to compare antioxidant activity of rosmarinic acid with that of the other mentioned phenolics, contradictory results are found. Thus, in the methyl-linoleate test (14), the antioxidant efficiency increases in the following order: chlorogenic acid, protocatechuic acid, caffeic acid, rosmarinic acid. However, using the method involving the free radical 2,2-Diphenyl-1-picrylhydrazyl (DPPH·) (15) the increasing order of antioxidative efficiency is rosmarinic acid, protocatechuic acid, caffeic acid. These data illustrate the above mentioned difficulty of comparing results obtained by application of different experimental methods.

Iron-chelating activity of rosmarinic acid

Superoxide anion is a relatively nonreactive species in aqueous solution, but in the presence of hydrogen peroxide and a transition metal such as iron, the extremely reactive hydroxyl radical may be generated through a superoxide anion-driven, metal-catalysed Fenton reaction. The hydroxyl radical is able to initiate lipid peroxidation directly through abstraction of hydrogen from fatty acids leading to the formation of off-flavors and other undesirable compounds.

One possible strategy for minimizing peroxidative damage consists in removing traces of heavy metals by the use of chelators. In this way, rosmarinic acid seems to be a good candidate for chelating iron and other metals since it bears two catechol and a carboxyl groups on its molecule.

When rosmarinic acid is incubated in the presence of iron ions, the formation of a rosmarinic acid-iron complex takes place. The formation of this complex can

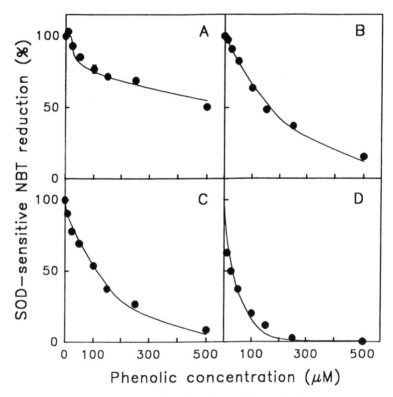

Figure 5. Effect of protocatechuic acid (A), caffeic acid (B), chlorogenic acid (C), and rosmarinic acid (D) concentrations on superoxide anions generated in a PMS/NADH system and determined by the superoxide dismutase-sensitive NBT reduction. Bars show standard errors.

Table II. Superoxide anion-scavenging efficiency and structure of the phenolic acids tested

Compound	IC_{50} (μM)	Structure
Protocatechuic acid	500	
Caffeic acid	157	
Chlorogenic acid	114	
Rosmarinic acid	33	

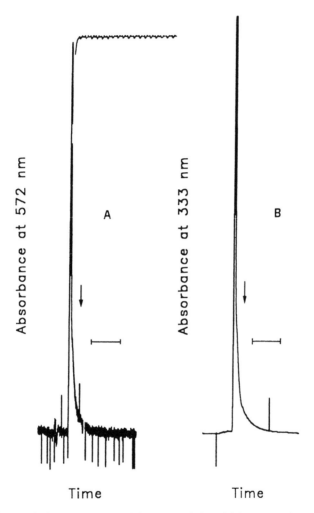

Figure 6. HPLC chromatograms of the rosmarinic acid-iron complex monitored at 572 nm (A) and at 333 nm (B). Arrows indicate retention time for rosmarinic acid (Bar = 4 minutes).

be followed by VIS spectroscopy and it presents an absorption maximum at 572 nm (*10*).

Titration curves for rosmarinic acid in the presence of different concentrations of iron ions indicate a stoichiometry of 1 to 1 for the complex. These data were confirmed by elemental analysis and inductively coupled plasma emission spectrometry (*10*). This 1 to 1 complex seems to be the only absorbent species in the reaction media. Support of this assertion come from graphical analysis of the consecutive spectra according to Coleman *et al.* (*16*) and from HPLC analysis of the reaction media (Figure 6). From this figure, it can also be deduced that, under the assay conditions, almost all rosmarinic acid is forming part of the complex, since absorbance at 333 nm for the expected retention time of rosmarinic acid is very low.

Although chelating activity of compounds bearing catechol groups has been largely known, there is no report on complexing properties of rosmarinic acid towards iron ions. In 1985 Banthorpe *et al.*, working with callus cultures of *Lavandula angustifolia*, demonstrated the secretion into the supporting agar of the (*Z,E*)-2-(3,4-dihydroxyphenyl)ethenyl ester of 3-(3,4-dihydroxyphenyl)-2-propenoic acid and of its (*E,E*)-isomer (*17*). These compounds, which are probably related biosynthetically with rosmarinic acid (*18*), are able to chelate Fe^{2+} ions forming intensely blue pigments (*17*). Since lavandin cell cultures also display blue pigmentation in the supporting agar and rosmarinic acid has been found to be the main phenolic constituent present in extracts of the medium (*19*), a participation of rosmarinic acid in the pigmentation of the culture media through the formation of complexes with iron ions cannot be discarded (*19*).

Conclusions

Rosmarinic acid fulfils the requirements for being considered as a potent antioxidant since it is not only capable of efficiently scavenging superoxide anions, but is also able to chelate iron ions. Studies on other possible mechanisms for rosmarinic acid acting as an antioxidant, such as hydrogen peroxide detoxification through peroxidase-catalyzed reactions (*8*), are now in progress. In the search for sources of this compound, lavandin cell cultures must be considered since they accumulate relatively high amounts of easy-to-purify rosmarinic acid.

Acknowledgments

This work has been partially supported by grants fron the CICYT (Spain), Projects # ALI 93/573 and ALI 95/1018.

Literature cited

1. Madsen, H. L.; Bertelsen, G. *Trends Food Sci. Technol.* **1995,** *6*, 271.
2. Maruta, Y.; Kawabata, J.; Niki, R. *J. Agric. Food Chem.* **1995,** *43*, 2592.
3. Lamaison, J. L.; Petitjean-Freytet, C.; Carnat, A. *Ann. Pharm. Fr.* **1990,** *48*, 103.
4. Petersen, M.; Häusler, E.; Meinhard, J.; Karwatzki, B.; Gertlowski, C. *Plant Cell Tiss. Org. Cult.* **1994,** *38*, 171.

5. De-Eknamkul, W.; Ellis, B. E. In *Biotechnology in Agriculture and Forestry*; Bajaj, Y. P. S., Ed.; Medicinal and Aromatic Plants I; Springer-Verlag: Berlin, 1988, Vol. 4; pp 310-329.

6. Whitaker, R. J.; Hashimoto, T.; Evans, D. A. *Ann. NY Acad. Sci.* **1984,** *435,* 364.

7. De-Eknamkul, W.; Ellis, B. E. *Planta Med.* **1984,** *51,* 346.

8. López-Arnaldos, T.; López-Serrano, M.; Ros Barceló, A.; Calderón, A. A.; Zapata, J. M. *Biochem. Molec. Biol. Int.* **1994,** *34,* 809.

9. Häusler, E.; Petersen, M.; Alfermann, W. *Plant Cell Rep.* **1993,** *12,* 510.

10. López-Arnaldos, T.; López-Serrano, M.; Ros Barceló, A.; Calderón, A. A.; Zapata, J. M. *Fresenius J. Anal. Chem.* **1995,** *351,* 311.

11. Economou, K. D.; Oreopoulou, V.; Thomopoulos, C. D. *J. Am. Oil Chem. Soc.* **1991,** *68,* 109.

12. Frankel, E. N. *Trends Food Sci. Technol.* **1993,** *4,* 220.

13. Shahidi, F.; Wanasundara, P. K. J. P. D. *Crit. Rev. Food Sci. Nutr.* **1992,** *32,* 67.

14. Cuvelier, M. E.; Richard, H.; Berset, C. *Biosci. Biotech. Biochem.* **1992,** *56,* 324.

15. Brand-Williams, W.; Cuvelier, M. E.; Berset, C. *Lebensm. Wiss. Technol.* **1995,** *28,* 25.

16. Coleman, J. S.; Varga, L. P.; Mastin, S. H. *Inorg. Chem.* **1970,** *9,* 1015.

17. Banthorpe, D. V.; Bilyard, H. J.; Watson, D. G. *Phytochemistry* **1985,** *24,* 2677.

18. Banthorpe, D. V.; Bilyard, H. J.; Brown, G. D. *Phytochemistry* **1989,** *28,* 2109.

19. López-Arnaldos, T.; López-Serrano, M.; Ros Barceló, A.; Calderón, A. A.; Zapata, J. M. *Nat. Prod. Lett.* **1995,** *7,* 169.

Chapter 18

Anti-inflammatory Antioxidants from Tropical Zingiberaceae Plants

Isolation and Synthesis of New Curcuminoids

Toshiya Masuda

Faculty of Integrated Arts and Sciences, University of Tokushima,
Minamijosanjima 1–1, Tokushima 770, Japan

Antiinflammatory active antioxidants were isolated from tropical
Zingiberaceae plant rhizomes. Their chemical structures were
determined to be new curcuminoids. In particular, the compounds
from *Zingiber cassumunar* (cassumunins and cassumunarins) have
complex curcuminoid structures and both groups have stronger
antioxidant and antiinflammatory activities than those of curcumin.
The total synthesis of one of these complex curcuminoids,
cassumunin A, was accomplished in 10 steps starting from *o*-
vanillin. The regio-specific effect of the hydroxyl and methoxy
groups on the antioxidant activity of curcuminoid was investigated
using synthetic curcumin analogs.

Natural antioxidants are important materials for the oxidative deterioration of
food, and some antioxidants also have potential prevention activity in the initial
stages of oxidation-related diseases. Plants used for food, including spice, are well
known to be rich in phenols. The phenols have been targeted as natural
antioxidants because most phenolic compounds have a radical trapping property
which produces an antioxidant activity (1). There are about fourteen hundred
species of the Zingiberaceae plants in the world. Most of these grow or are
cultivated in tropical and subtropical Asia (2). The Zingiberaceae plants typically
have large rhizomes. Asian people have used these for food and spice, and also for
traditional medicines to maintain health (3). In connection with our research to
discover new natural antioxidants, which are beneficial for human health, we have
been investigating tropical Zingiberaceae plants as a source for new biologically
active antioxidants.

Antioxidant Activity of Tropical Zingiberaceae Plant Rhizomes

We have examined the antioxidant activity of nine cultivated species of
Zingiberaceae rhizomes which were collected in Indonesia and Okinawa. The
antioxidant activity was judged by the inhibition of linoleic acid oxidation in an

ethanol-buffer system. The results obtained using the thiocyanate and TBA methods (4) are shown in Figure 1. The antioxidant activity of the acetone extract of the rhizomes increased in the order *Curcuma heyneana* < *Phaeomeria speciosa* < *Curcuma aeruginosa* < *Amomum kepulaga* < *Curcuma mangga* < *Zingiber cassumunar* < *Curcuma xanthorrhiza* < *Alpinia galanga* < *Curcuma domestica* using the thiocyanate method, and in the order *Curcuma heyneana* < *Phaeomeria speciosa* < *Curcuma aeruginosa* = *Curcuma mangga* < *Amomum kepulaga* < *Zingiber cassumunar* < *Curcuma xanthorrhiza* < *Alpinia galanga* < *Curcuma domestica* using the TBA method.

Curcumin is a typical constituent of tropical Zingiberaceae plant (5) and has strong antioxidant activity (6). We analyzed the quantity of curcumin and two known analogs in the acetone extract of the rhizomes. Although the extracts from *Curcuma domestica*, *Curcuma xanthorrhiza*, and *Zingiber cassumunar* were determined to contain a relatively high amount of curcumin by HPLC analysis, their strong antioxidant activity could not be explained only by the curcuminoid contents (7). The results indicated the presence of new additional antioxidants in the rhizomes.

New Antiinflammatory Antioxidants from Tropical Zingiberaceae Plants

Recently, curcumin has received much attention because of its interesting biological activity, which may be related to its antioxidant property (8). Antiinflammatory activity is an important biological activity of curcumin (9), because inflammation is one of the peroxidation-related events in living organisms and its inhibition suggests that curcumin works as an antioxidant even in living cells. Huang *et al.* reported antitumor promotion activity of curcumin and clarified that its activity is linked with the suppression of arachidonic acid metabolism (10). This metabolic pathway is one of the lipid peroxidation events in living organisms. Inflammation is also well known to be closely related to the metabolism of arachidonic acid. We have been investigating antiinflammatory active antioxidants in tropical Zingiberaceae plants and succeeded in isolating such antioxidants from three species of tropical Zingiberaceae plants.

Curcuma domestica. Half-processed rhizomes of *Curcuma domestica* are known as turmeric and are used as a food coloring agent. The rhizomes are also used as traditional medicines. New antioxidants (compounds **2** and **3**) were isolated from the dry rhizomes of the plant, the structures of which are shown in Figure 2. Antioxidant activity of the compounds are displayed in Figure 3 (11).

Curcuma xanthorrhiza. *Curcuma xanthorrhiza* is a medicinal ginger in Indonesia, however, the rhizomes are sometimes used for spice as a substitute for *Curcuma domestica*. We isolated new antioxidants (compounds **4** and **5**) from the rhizomes. Their structures and antioxidant activity are presented in Figures 2 and 4, respectively (12).

Zingiber cassumunar. *Zingiber cassumunar* has large yellow rhizomes similar to *Curcuma domestica*. The rhizomes are known to have antiinflammatory activity (13) and are used as traditional medicines in Thailand and Indonesia. We have succeeded in isolating new antioxidants with antiinflammatory activity from the rhizomes based on both the antioxidant and antiinflammatory assays (14, 15). Figure 2 shows the structures of six new compounds |cassumunins A-C (16) (compounds **6**, **7**, and **8**, respectively) and cassumunarins A-C (17) (compounds **9**, **10**, and **11**, respectively)| isolated along with curcumin and 5'-methoxycurcumin. The antioxidant activity of the cassumunins and cassumunarins is also shown in Figures 5 and 6, respectively.

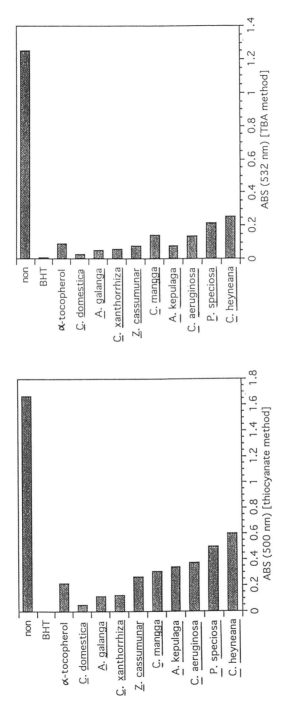

Figure 1. Inhibitory Activity of Tropical Zingiberaceae Extracts (0.02% w/v) against Autoxidation of Linoleic acid (16 mM) in an Ethanol-buffer (pH 7.0) System.

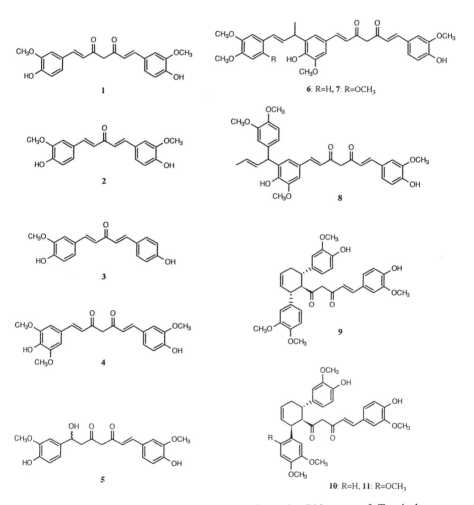

Figure 2. The Structures of Antioxidants from the Rhizomes of Tropical Zingiberaceae Plants

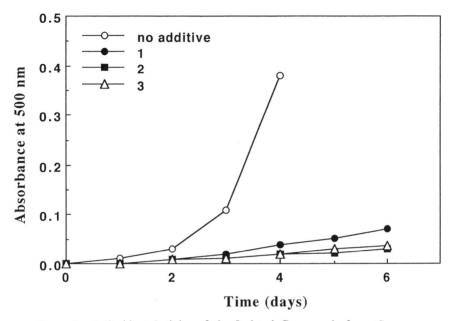

Figure 3. Antioxidant Activity of the Isolated Compounds from *Curcuma domestica*.

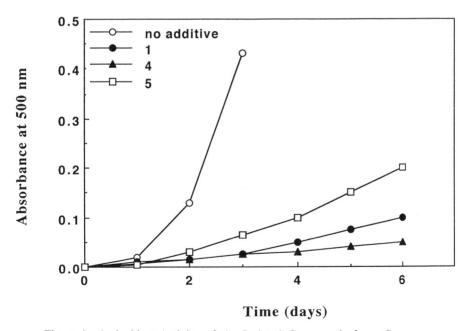

Figure 4. Antioxidant Activity of the Isolated Compounds from *Curcuma xanthorrhiza*.

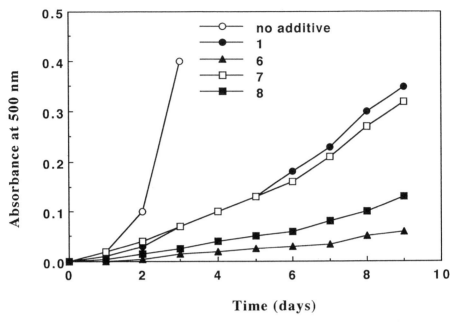

Figure 5. Antioxidant Activity of Cassumunins A (**6**), B (**7**), and C (**8**).

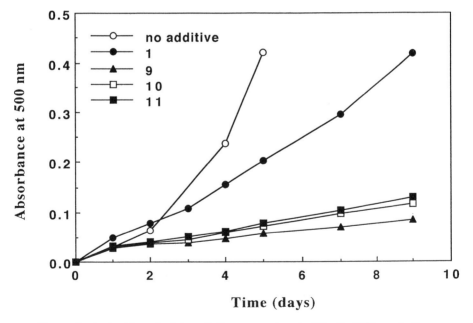

Figure 6. Antioxidant Activity of Cassumunarins A (**9**), B (**10**), and C (**11**).

Antiinflammatory Activity of the Isolated Antioxidants. Newly isolated compounds from the three Zingiberaceae plants previously mentioned are structurally related to curcumin. We examined their antiinflammatory activity (18). Curcumin (0.6 μmol) showed inhibitory activity of edema formation, which is caused by an inflammation inducer, TPA (12-*O*-tetradecanoylphorbol-13-acetate, 2 μg), on mouse ear. The activity of the isolated compounds using the mouse ear method is summarized in Table I.

Synthesis of Cassumunin A, a Complex Curcuminoid from *Zingiber cassumunar*

We have begun the total synthesis of several new complex curcuminoids and succeeded in the synthesis of cassumunin A, the most potent antiinflammatory antioxidant from *Zingiber cassumunar*. The synthesis of curcumin was achieved by many groups (19), however, the most efficient method was first reported by Pabon (20), which involves the Knoevenagel condensation of the boron complex of 2,4-pentadione with vanillin. In the synthesis strategy, we employed this method for the coupling of two different types of benzaldehyde with 2,4-pendadione.

Methoxymethylated *o*-vanillin **12**, which was prepared from *o*-vanillin and methoxymethylchloride, was condensed with 3',4'-dimethoxyacetophenone, giving the conjugated ketone **13** in 97% yield. One methyl group was introduced at the β-position of the carbonyl group in **13** by treatment of $(CH_3)_2CuLi$ in 94% yield (compound **14**). The methoxymethyl group of **14** was removed by acid treatment (1N HCl) in methanol to give **15** in 76% yield. The carbonyl group of **15** was reduced to a hydroxyl group by $NaBH_4$ in 98% yield (compound **16**). After diacetylation of **16**, the diacetate **17** was heated at 160 °C in DMSO to introduce a double bond at the appropriate position in 90% overall yield (compound **18**) from **15**. After hydrolysis of the acetyl group of **18** (compound **19**), regioselective introduction of aldehyde function at the 4-position was achieved by treatment with chloroform under alkaline conditions, giving aldehyde **20** in 40% yield. The coupling reaction of **20** with diketone **21** under Pabon's conditions (20) gave cassumunarin A in 41% yield. The synthetic cassumunin A was identical with the natural compound based on spectroscopic methods (Figure 7).

Synthesis of New Curcumin Analogs and Their Antioxidant and Antiinflammatory Activities

Among plant constituents, there are many kinds of phenolic compounds, such as flavonoids, coumarins, and phenol carboxylic acids. It is well known that they have various regioisomers in positions of the hydroxyl and methoxyl group substituents. The structure and antioxidant efficiency relationships for tocopherols (21) and flavonoids (22) have been well investigated, however, little is known about curcuminoids. We have been interested in the regio-effect of the hydroxyl and methoxyl groups in the benzene rings of curcumin on their antioxidant and antiinflammatory activities.

We synthesized eleven curcumin analogs as shown in Figure 8, and examined their antioxidant activity using linoleic acid as a substrate and AIBN as an oxidation inducer. The data are shown in Figures 9-12 (23). Figure 9 shows that the hydroxyl group at 2 or 4-positions is important for antioxidant activity. Figures 10-12 show that a methoxyl group at the *ortho* or *para*-positions increases the activity. The 4-hydroxy-3-methoxy compound **31** (curcumin), the 2-hydroxy-3-methoxy compound **26**, and the 2-hydroxy-5-methoxy compound **28** show the

**Table I. Inhibitory Activity of the Isolated Compounds (0.6 μmol)
on TPA (2 μg)-Induced Mouse Ear Inflammation**

Experiment	Treatment Left Ear	Right Ear	n^a	$D^b \pm SE(mg)$	Inhibition(%)	Significance
1.						
		TPA	10	16.6±0.9		
	1+TPA	TPA	10	9.7±1.1	58	$p<0.01$
	2+TPA	TPA	5	8.7±1.4	52	$p<0.01$
	3+TPA	TPA	5	5.3±0.4	32	$p<0.01$
	4+TPA	TPA	6	11.5±1.3	69	$p<0.01$
2.						
		TPA	6	17.5±0.9		
	1+TPA	TPA	5	8.9±0.9	51	$p<0.01$
	6+TPA	TPA	5	14.5±1.0	83	$p<0.01$
	7+TPA	TPA	5	13.3±1.2	76	$p<0.01$
	8+TPA	TPA	5	13.1±1.3	75	$p<0.01$
3.						
		TPA	6	16.5±0.6		
	1+TPA	TPA	5	9.2±1.3	56	$p<0.01$
	9+TPA	TPA	5	12.3±2.1	75	$p<0.01$
	10+TPA	TPA	5	10.3±1.4	62	$p<0.01$
	11+TPA	TPA	5	10.3±0.8	62	$p<0.01$

[a] n: number of mice, [b] D: mean of weight differnce between right ear and left ear.

**Table II. Inhibitory Activity of Curcuminoids (0.6 μmol)
on TPA (2 μg)-Induced Mouse Ear Inflammation**

Treatment Left Ear	Right Ear	n^a	$D^b \pm SE$	Inhibition(%)	Significance
	TPA	4	17.4±0.9		
31(1)+TPA	TPA	4	9.7±1.9	56	$p<0.01$
28+TPA	TPA	4	10.0±1.2	58	$p<0.01$
26+TPA	TPA	4	4.4±1.5	25	$p<0.05$

[a] n: number of mice, [b] D: mean of weight differnce between right ear and left ear.

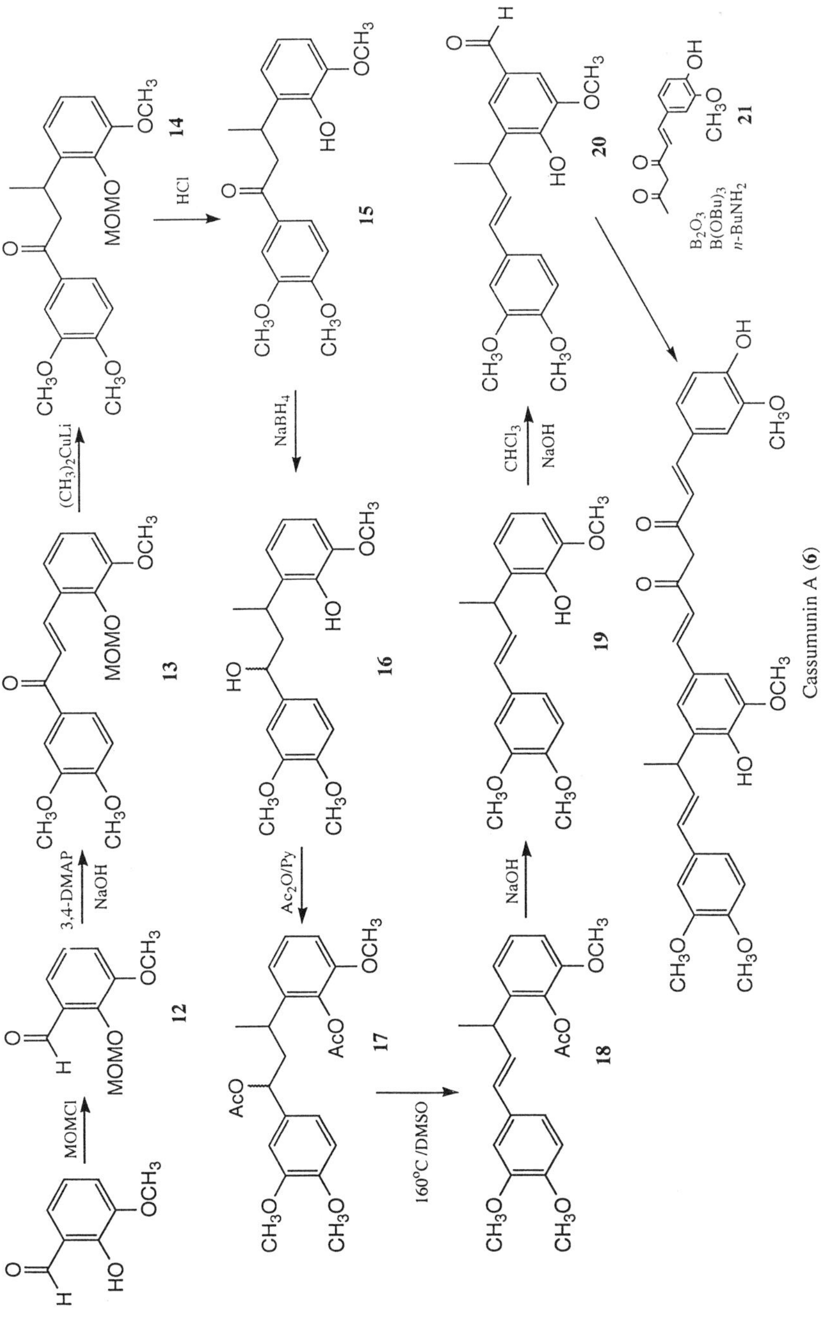

Figure 7.   Synthetic Route for Cassumunin A.

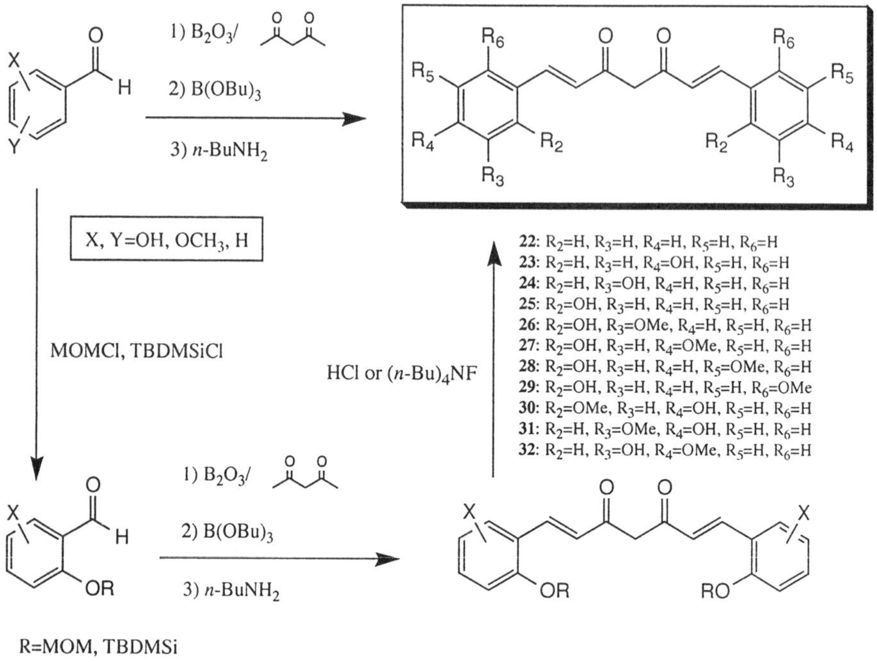

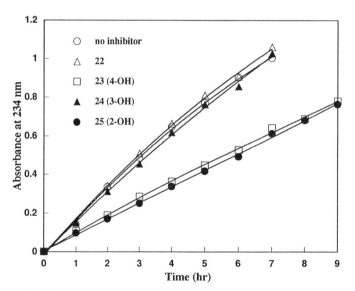

1) B_2O_3/
2) $B(OBu)_3$
3) n-$BuNH_2$
X, Y=OH, OCH_3, H
MOMCl, TBDMSiCl
HCl or $(n\text{-}Bu)_4NF$
22: R_2=H, R_3=H, R_4=H, R_5=H, R_6=H
23: R_2=H, R_3=H, R_4=OH, R_5=H, R_6=H
24: R_2=H, R_3=OH, R_4=H, R_5=H, R_6=H
25: R_2=OH, R_3=H, R_4=H, R_5=H, R_6=H
26: R_2=OH, R_3=OMe, R_4=H, R_5=H, R_6=H
27: R_2=OH, R_3=H, R_4=OMe, R_5=H, R_6=H
28: R_2=OH, R_3=H, R_4=H, R_5=OMe, R_6=H
29: R_2=OH, R_3=H, R_4=H, R_5=H, R_6=OMe
30: R_2=OMe, R_3=H, R_4=OH, R_5=H, R_6=H
31: R_2=H, R_3=OMe, R_4=OH, R_5=H, R_6=H
32: R_2=H, R_3=OH, R_4=OMe, R_5=H, R_6=H
R=MOM, TBDMSi

Figure 8. Structures and Synthesis of Curcuminoids (**23-32**).

no inhibitor
22
23 (4-OH)
24 (3-OH)
25 (2-OH)
Absorbance at 234 nm
Time (hr)

Figure 9. Inhibitory Activity of Curcuminoids (**22-25**) against AIBN-initiated Linoleic Acid Oxidation (Linoleic Acid: 54 mM, AIBN: 81 mM, Curcuminoid: 170 μM).

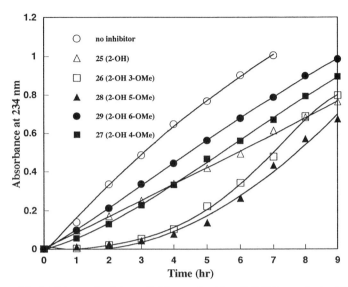

Figure10. Inhibitory Activity of Curcuminoids (**25-29**) against AIBN-initiated Linoleic Acid Oxidation (Linoleic Acid: 54 mM, AIBN: 81 mM, Curcuminoid: 170 μM).

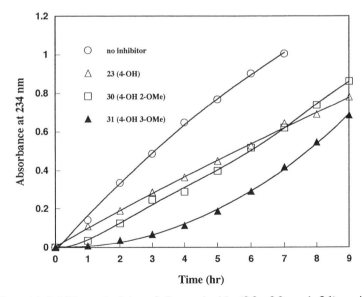

Figure11. Inhibitory Activity of Curcuminoids (**23**, **30**, and **31**) against AIBN-initiated Linoleic Acid Oxidation (Linoleic Acid: 54 mM, AIBN: 81 mM, Curcuminoid: 170 μM).

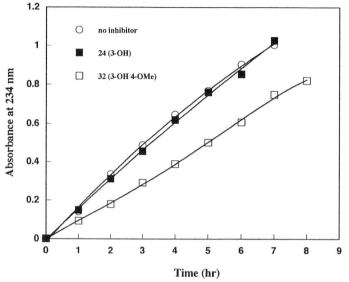

Figure12. Inhibitory Activity of Curcuminoids (**2 4** and **3 2**) against AIBN-initiated Linoleic Acid Oxidation (Linoleic Acid: 54 mM, AIBN: 81 mM, Curcuminoid: 170 μM).

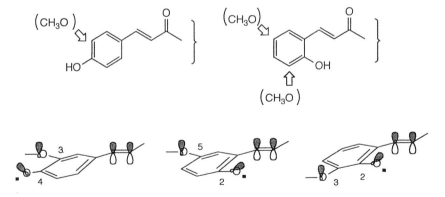

Figure13. Stabilization Effect of Phenolic Radical by Methoxyl Oxygen and Olefin of Curcuminoids (**2 6**, **2 8**, and **3 1**).

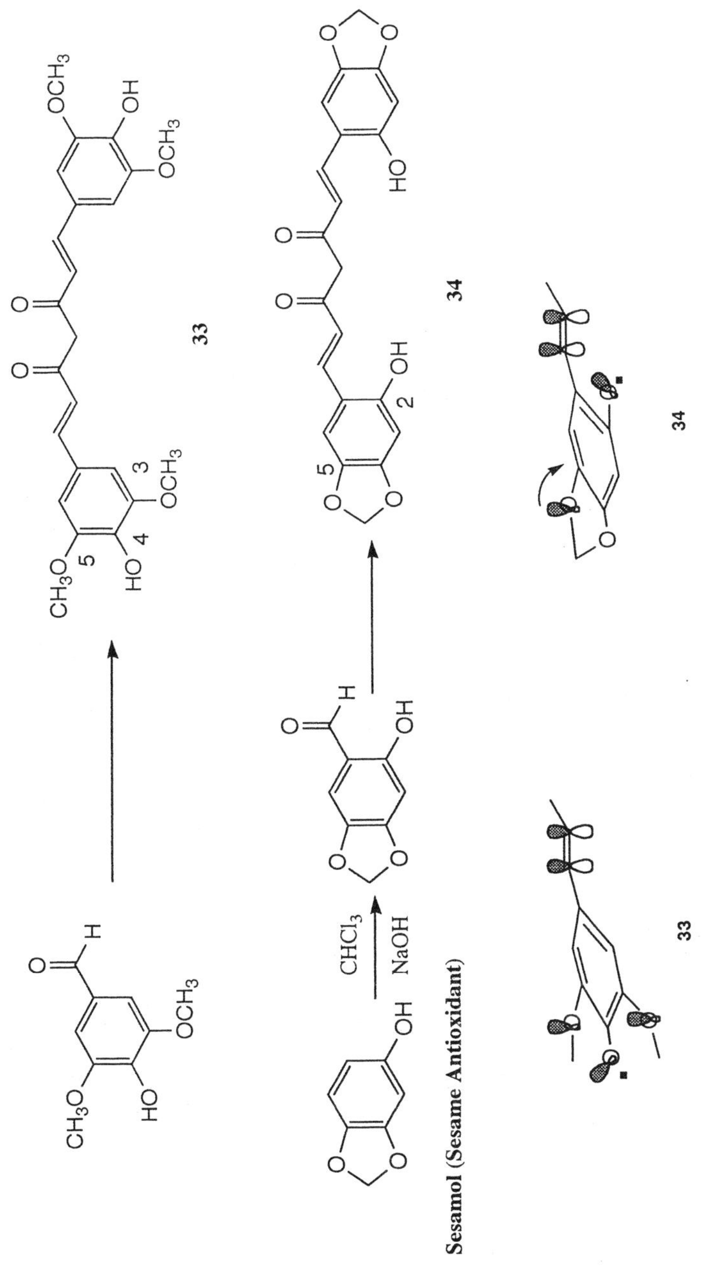

Figure 14. Synthesis of Curcuminoids (**33** and **34**).

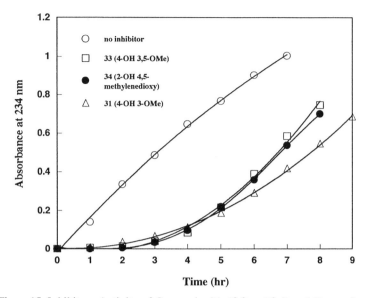

Figure 15. Inhibitory Activity of Curcuminoids (**3 3** and **3 4**) and Curcumin (**3 1**) against AIBN-initiated Linoleic Acid Oxidation (Linoleic Acid: 54 mM, AIBN: 81 mM, Curcuminoid: 170 mM).

strongest activity in this assay system and their efficiency is almost equivalent. These data indicate that the following factors are important for the curcuminoid's antioxidant activity; 1) the phenolic radical can be conjugated to the double bond in the alkyl part at the 1-position. 2) the phenolic radical can be stabilized by the lone pair of the methoxyl oxygen, as shown in Figure 13. These parameters are similar to those in tocopherol's case (24). We synthesized two curcumin analogs, the 4-hydroxy-2,3-dimethoxy analog **33** and the 2-hydroxy-3,4-methylenedioxy analog **34** (Figure 14). In the former compound the phenolic radical can be stabilized by the two methoxyl oxygens and the radical of the latter compound can be stabilized by one of the methylendioxyl oxygens, whose lone pair is well overlapped with the phenolic radical because the five membered ring restricts the lone pair obital of the oxygen in a good direction for the delocalization of the phenolic radical to the oxygen (25).They showed slightly stronger antioxidant activity than that of curcumin (Figure 15).

The antiinflammatory activity of the two analogs (compounds **26** and **28**), which showed equivalent antioxidant activity to curcumin, are summarized in Table II. These analogs have antiinflammatory activity similar to curcumin, indicating that the regio-specificity of substituents on the aromatic rings to the antiinflammatory activity may not be so important and only the antioxidant efficiency of the substance may important. Further studies along these lines are now in progress.

Acknowledgments

This work was supported in part by grants from the Uehara Memorial Foundation, Takano Research Foundation, and Ministry of Education, Science and Culture of Japan. The author thanks Dr. A. Jitoe, Ms. J. Isobe, and Ms. A. Kida of Osaka City University for their work in this project. The author also thanks Dr. Tom J.

Mabry of the University of Texas at Austin for our collaboration in Texas, and Dr. N. Nakatani of Osaka City University and Dr. S. Yonemori of the Univ. of Ryukyus for their support of plant collection.

References

1. Osawa, T.; Ramaratham, S.; Kawakishi, S.; Namiki, M., *Phenolic Compounds in Food and Their Effects on Health II,* ACS Symposium Series 507; American Chemical Society; Washington, DC, 1992, Chapter 10, pp 122-134.
2. Hegnauer, R., *Chemotaxonomie der Pflanzen,* Birkhauser Verlag; Basel, 1963, Vol. 2.; Chapter 48, pp 451-471.
3. Corner, E. J. H.; Watanabe, K., *Illustrated Guide to Tropical Plants,* Hirokawa Publishing; Tokyo, 1969, pp 1069-1086.
4. Osawa, T.; Namiki, M., *Agric. Biol. Chem.*, **1981**, *45*, 735-739.
5. Kuroyanagi, M.; Natori, S., *Yakugaku Zasshi,* **1970**, *90*, 1467-1470.
6. Toda, S.; Miyase, T.; Arichi, H., Tanizawa, H.; Takino, Y., *Chem. Phram. Bull.,* **1985**, *33*, 1725-1728.
7. Jitoe, A.; Masuda, T.; Tengah, I. G. P.; Suprapata, D. N.; Gara, I. W.; Nakatani, N., *J. Agric. Food Chem.,* **1992**, *40*, 1337-1340.
8. Ammon, H. P. T.; Wahl, M. A., *Planta Med.,* **1991**, *57*, 1-7.
9. Rao, T. S.; Basu, N.; Siddiqui, H. H., *Indian J. Med. Res.,* **1982**, *75*, 574-578.
10. Huang, M.-T.; Smart, R. C.; Wong, C.-Q.; Conney, A. H., *Cancer Res.,* **1991**, *51*, 813-819.
11. Masuda, T.; Jitoe. A.; Isobe, J.; Nakatani, N., *Phytochemistry,* **1993**, *32*, 1557-1560.
12. Masuda, T., Isobe, J.; Jitoe, A.; Nakatani, N., *Phytochemistry,* **1992**, *31*, 3645-3647.
13. Ponglux, D.; Wongseripipatana, S.; Phadungcharoen, T.; Ruangrungsri, N.; Likhiwitayawuid, K., *Medicinal Plants;* Victory Point; Bangkok, **1987**, pp 275.
14. Masuda, T.; Jitoe, A., *J. Agric. Food Chem.,* **1994**, *42*, 1850-1856.
15. Masuda, T.; Jitoe, A.; Mabry, T. J., *J. Am. Oil Chem. Soc.,* **1995**, *72*, 1053-1057.
16. Masuda, T.; Jitoe, A.; Nakatani, N., *Chemistry Lett.,* **1993**, 189-192.
17. Jitoe, A.; Masuda, T.; Mabry, T. J., *Tetrahedron Lett.,* **1994,** *35,* 981-984.
18. Gshwendt, M.; Kittstein, K., Furstenberger, G.; Marks, F., *Cancer Lett.,* **1984,** *25*, 177-187.
19. Arrieta. A. F., *Pharmazie,* **1993**, *48,* 696-697.
20. Pabon, H. J. J., *Recuil,* **1964**, *83*, 379-385.
21. Barclay, L. R. C.; Edwards, C. D.; Mukai, K.; Egawa, Y., Nishi, T., *J. Org. Chem.,* **1995**, *60*, 2739-2744.
22. Mora, A.; Rios, P. J. L.; Alcaraz, M. J.; *Biochem. Pharmacol.,* **1990,** *40*, 793-797.
23. Masuda, T; Kida, A., Proceeding of Annual Meeting of Japanese Society of Nutrition and Food Science, **1995**, pp 103.
24. Hughes, L.; Burton, G. W.; Ingold, K. U.; Slaby, M.; Foster, D. O., *Phenolic Compounds in Food and Their Effects on Health II,* ACS Symposium Series 507; American Chemical Society; Washington, DC, 1992, Chapter 14, pp 184-199.
25. Burton, G. W.; Ingold, K. U., *Acc. Chem. Res.,* **1986**, *19*, 194-201.

Chapter 19

Curcumin: A Pulse Radiolysis Investigation of the Radical in Micellar Systems

A Model for Behavior as a Biological Antioxidant in Both Hydrophobic and Hydrophilic Environments

A. A. Gorman[1], I. Hamblett[1], T. J. Hill[1], H. Jones[1],
V. S. Srinivasan[2], and P. D. Wood[1]

[1]Department of Chemistry, University of Manchester,
Manchester M13 9PL, United Kingdom
[2]Center for Photochemical Sciences, Bowling Green State University,
Bowling Green, OH 43403

Curcumin has been solubilised in aqueous anionic, cationic and neutral micellar systems and in aqueous buffer. The formation of its radical has been observed in all situations. The radical is effectively inert to oxygen and reacts with the natural antioxidants vitamin C and vitamin E and Trolox to a degree dependent on location. It is concluded that the anticarcinogenic activity of curcumin may well be a consequence of its ability to access both hydrophobic and hydrophilic environments, act as a radical scavenger and be repaired by both vitamin C and vitamin E.

Curcumin (1), the major constituent of the spice turmeric, is a molecule of considerable interest as a consequence of its known biological activity. This includes the light-induced oxidative killing of bacteria (1,2) and anticarcinogenesis related to inhibition of lipid peroxidation (3-6). As part of a program aimed at understanding the anticarcinogenic activity of this molecule (cf. Ref. 7) we have examined the behavior of

1

curcumin in micellar systems with emphasis on its location, the formation of the radical and the reaction of that radical with oxygen and key antioxidants. The surfactants employed were sodium dodecyl sulfate (SDS), cetyl trimethylammonium bromide (CTAB) and Triton X-100 (TX) with anionic, cationic and neutral head groups respectively. The curcumin radical was formed by the standard pulse radiolytic sequence (8,9) summarised in equations 1-5 where curcumin (Cur-OH) donates an electron to the azide radical prior to loss of a proton.

$$H_2O + e \rightarrow H^\bullet + OH^\bullet + e(\text{solvated}) \tag{1}$$

$$e(\text{solvated}) + N_2O + H_2O \rightarrow N_2 + OH^- + OH^\bullet \tag{2}$$

$$OH^\bullet + N_3^- \rightarrow OH^- + N_3^\bullet \tag{3}$$

$$N_3^\bullet + \text{Cur-OH} \rightarrow N_3^- + \text{Cur-OH}^{\bullet+} \tag{4}$$

$$\text{Cur-OH}^{\bullet+} \rightarrow \text{Cur-O}^\bullet \tag{5}$$

Experimental

Pulse radiolysis experiments were essentially as previously described (*10*). Digitized data were transferred to the memory of a DAN 486 DX 33 PC for analysis. Water was doubly-distilled from potassium permanganate. Curcumin was as described (*7*). Cetyl trimethylammonium bromide (Fisons), sodium azide (BDH), sodium dodecyl sulfate (Aldrich), Triton X-100 (Aldrich), Trolox (Aldrich), vitamin C (Sigma) and vitamin E (Aldrich) were used as received. Solutions were bubbled with either nitrous oxide (British Oxygen Company) or nitrous oxide containing 10% oxygen (Air Products). Rate constants determined by time-resolved techniques are accurate to ± 5%.

Results.

Curcumin was readily solubilised in aqueous surfactant systems in which the surfactant concentration was above the critical micelle concentration (cmc). In each case the electronic absorption spectrum was independent of surfactant concentration above the cmc.

Absorption Spectra. The electronic absorption spectra of curcumin solubilised in aqueous SDS and TX (Figure 1a) systems were essentially identical ($\lambda_{max} \sim 417$ nm) and absolutely typical of the corresponding spectra in hydrocarbon solvents such as cyclohexane, a clear indication that, in such systems, a curcumin molecule on average occupies a hydrophobic region of the micellar system, i.e. is "within" a micelle. In contrast, the spectrum of curcumin in the aqueous CTAB system (Figure 1b) was typical of the orange curcumin anion ($\lambda_{max} \sim 480$ nm) and it would appear that, in this system, curcumin is predominantly on the outer surface of the micelles, anionic, and associated with the tertiary ammonium head groups of the CTAB components. These conclusions are supported by pulse radiolysis experiments (*vide infra*).

Pulse Radiolytic Formation of the Radical in Micellar Systems. Pulse radiolysis of N_2O-bubbled aqueous, pH 7 buffered, micellar solutions of curcumin (2.0×10^{-5} mol l^{-1}) containing sodium azide (2.0×10^{-2} mol l^{-1}) led to formation of a species with $\lambda_{max} \sim 490$ nm. This is exemplified for the case of CTAB in Figure 2 which shows the formation of the 490 nm species and the bleaching of the curcumin ground state as a function of time. These processes take place on the same microsecond timescale (*cf*. Figures 2a and 2b) and the slow decay of the 490 nm species on millisecond timescales (Figure 2c) was second-order, as expected for a radical-radical combination process. There were only minor variations in the kinetics of radical growth and decay for the different micellar systems for a given curcumin concentration. Of key importance was the finding that use of N_2O containing 10% oxygen led to absolutely no change in the kinetics of decay of the radical absorption at 490 nm. This allows an upper limit of $\sim 1.0 \times 10^2$ l mol^{-1} s^{-1} to be placed on the rate constant for reaction with oxygen. It is clear that in these systems the curcumin radical is indeed being produced and is inert to oxygen on the timescales of its existence in our experiments. It must be recognised that in the real biological situation the radical would not decay via second-

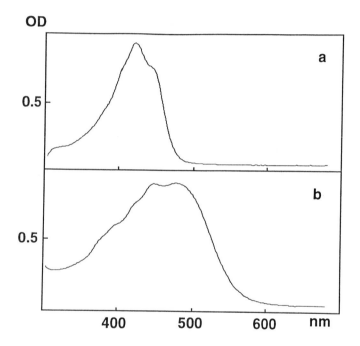

Figure 1. Electronic absorption spectra of curcumin (2.0×10^{-5} mol l^{-1}) solubilised in an aqueous solution of (a) TX (5.0×10^{-3} mol l^{-1}) and (b) CTAB (5.0×10^{-3} mol l^{-1}).

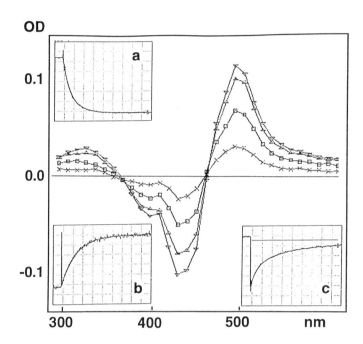

Figure 2. Transient absorption spectra measured 6.4 µs (X), 22 µs (□), 58 µs (Δ) and 126 µs (∇) after pulse radiolysis of a solution of curcumin (2.0 x 10⁻⁵ mol l⁻¹), CTAB (5.0 x 10⁻⁵ mol l⁻¹) and sodium azide (2.0 x 10⁻² mol l⁻¹) in N₂O-saturated pH 7 buffer. Insets: time dependence of (a) transient formation at 490 nm with first-order fit, $k' = 1.0$ x 10^5 s⁻¹, 4.6 % absorption/division, 10 µs/division, (b) corresponding ground-state bleaching at 430 nm with first-order fit, $k' = 0.71$ x 10^5 s⁻¹, 6.7 % absorption/division, 10 µs/division and (c) transient decay at 490 nm with second-order fit, 4.2 % absorption/division, 50 ms/division.

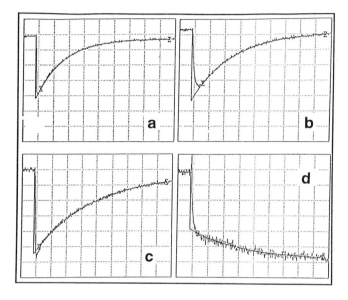

Figure 3. (a) and (b): Repeat of experiments corresponding to Figure 2 showing transient decay at 490 nm in the presence of (a) vitamin C (2.0×10^{-5} mol l^{-1}) with first-order fit, $k' = 2.0 \times 10^3$ s^{-1}, 4.9 % absorption/division, 300 μs/division and (b) Trolox (2.0×10^{-5} mol l^{-1}) with first-order fit, $k' = 8.5 \times 10^3$ s^{-1}, 4.6 % absorption / division, 50 μs/division. (c) and (d): Repeat of experiments corresponding to Figure 2 using TX (5.0×10^{-3} mol l^{-1}) in the presence of vitamin E (2.0×10^{-5} mol l^{-1}) showing (c) transient decay at 490 nm with first-order fit, $k' = 1.1 \times 10^3$ s^{-1}, 1.2 % absorption/division, 250 μs/division and (b) corresponding growth of absorption at 430 nm with first-order fit, $k' = 1.2 \times 10^3$ s^{-1}, 0.6 % absorption/division, 250 μs/division.

order combination with itself, but by a much slower pseudo first-order process involving ambient scavengers. Inefficient reaction with oxygen might become more important in that situation.

2

3   4

The next question to be addressed at this stage was whether the location of radical formation would influence its ability to react with additive molecules expected to be good radical scavengers. The potential scavengers chosen were vitamin E (2), vitamin C (3) and the water-soluble model of the former, Trolox (4). In the event of efficient repair of the curcumin radical one would anticipate firstly a significant shortening of the latter's lifetime and a change to pseudo first-order decay, if the quencher concentration was significantly higher than that of the radical. Thus, the above experiments were repeated in the presence of 2.0×10^{-5} mol l^{-1} scavenger. In all of these experiments the scavenger will of course compete with curcumin for the azide radical. The consequence of this is formation of both the curcumin radical and the potential scavenger radical on microsecond timescales (cf. Figure 2a). The subsequent interactions can be monitored as a consequence of the fact that the absorption maximum of the curcumin radical (490 nm) is significantly shifted from those of vitamin C (~ 360 nm) and vitamin E and Trolox (~ 430 nm).

Scavenging of the Curcumin Radical Produced in CTAB Micelles. In the presence of vitamin E the curcumin radical decayed via clean second-order kinetics on a timescale identical to that of Figure 2c, i.e. in this situation vitamin E, which must be solubilised within the micelles, is unable to scavenge the curcumin radical. Since subsequent experiments demonstrate repair of the latter by vitamin E, these data clearly agree with the proposal that the curcumin in this particular microheterogeneous environment occupies an extramicellar location as the corresponding mono-anion. Strong support for this conclusion came from the finding that the water-soluble vitamin C and Trolox both quenched the curcumin radical efficiently as shown in Figures 3a and 3b. The decay of the radical absorption followed clean first-order kinetics, corresponding to rate constants for reaction between the curcumin radical and the scavenger of 1.0×10^8 l mol^{-1} s^{-1} for vitamin C and 4.3×10^8 l mol^{-1} s^{-1} for Trolox. Although the latter is somewhat more efficient as a scavenger, the experimental rate constants are too close to allow significant conclusions to be drawn.

Scavenging of the Curcumin Radical in SDS and TX Micelles. In stark contrast to the situation for CTAB above, the curcumin radical produced in either SDS- or TX-based micellar systems was efficiently quenched by vitamin E solubilised within the micelles as exemplified for TX in Figures 3c and 3d. These show the decay of the

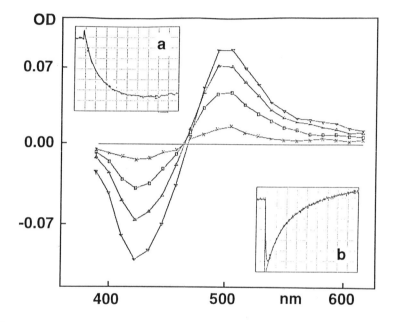

Figure 4. Transient absorption spectra measured 5.2 μs (X), 12 μs (□), 25 μs (Δ) and 58 μs (∇) after pulse radiolysis of a solution of curcumin (2.0 x 10^{-5} mol l^{-1}) and sodium azide (2.0 x 10^{-2} mol l^{-1}) in N$_2$O-saturated pH 7 buffer. Insets: time dependence of (a) transient formation at 490 nm with first-order fit, $k' = 7.2$ x 10^4 s^{-1}, 1.0 % absorption/division, 10 μs/division and (b) transient decay at 490 nm with second-order fit, 0.7 % absorption/division, 250 μs/division.

curcumin radical at 490 nm and the corresponding growth of the vitamin E radical at 430 nm on the same timescales. These data clearly indicate that the quenching of the curcumin radical by vitamin E does indeed involve repair and supports the earlier conclusion that in the SDS and TX systems the curcumin is solubilised within the hydrophobic region of the micelle. When these experiments were repeated in the presence of vitamin C (2.0×10^{-5} mol l^{-1}), the decay of the curcumin radical was only marginally faster than the blank and still largely second-order. In the presence of the same concentration of Trolox quenching occurred to an extent which made the decay predominantly first-order but the rate constant was an order of magnitude lower than in the CTAB-based system (*vide supra*). Clearly Trolox is more hydrocarbon-like than vitamin C and has a higher probability of accessing the hydrophobic regions.

Formation of the Curcumin Radical in Water. It was possible to produce 2.0×10^{-5} mol l^{-1} curcumin in water by dropping a solution in 0.1 M NaOH into a large excess of pH 7 buffer. That this was a real homogeneous solution, as opposed for instance to a microcrystalline dispersion, was clearly demonstrated by the spectrum which was essentially identical to that in other hydroxylic solvents such as ethanol and in particular the kinetics of its reaction with the azide radical, with and without addition of vitamin C and Trolox. Thus, pulse radiolysis of such a solution, N_2O-bubbled, containing sodium azide led to formation of the curcumin radical (Figure 4a) on timescales essentially identical to those for the CTAB-based micellar system for which curcumin is thought to be in the anionic form on the micellar surface. The spectral changes with time (Figure 4) were virtually identical to those for the micellar systems, particularly the SDS and TX systems where the curcumin is in the neutral form, and the radical decay was second-order as expected (Figure 4b).

Scavenging of the Curcumin Radical in Water. Addition of either vitamin C or Trolox (2.0×10^{-5} mol l^{-1} in each case) led to efficient quenching of the curcumin radical. Rate constants for scavenging were 1.2×10^8 l mol^{-1} s^{-1} and 1.6×10^8 l mol^{-1} s^{-1} for vitamin C and Trolox respectively. The raw data are exemplified in the case of Trolox in Figure 5. Insets a and b show the decay of the curcumin radical at 490 nm and the growth of the Trolox radical at 430 nm respectively. These occur on the same timescales. Inset b shows the initial 'immediate' formation of the Trolox radical as a consequence of competition for the azide radical. The spectra in Figure 5 clearly show the loss of the curcumin radical at 490 nm with a concomitant increase in the intensity of the Trolox radical at 430 nm. An isosbestic point close to 450 nm is apparent. Again these data confirm that the scavenging involves repair of the curcumin radical.

Conclusions.

It is clear from the experiments described that the molecule curcumin can access both hydrophobic and hydrophilic environments of a microheterogeneous system. In addition, the corresponding radical is stable with respect to oxygen, at least on our experimental timescales, but can be repaired by both the natural antioxidants vitamins C and E in hydrophilic and hydrophobic environments respectively. It is currently our view that the combination of amphiphilic character and radical stability/repair behavior referred to is the key to the importance of curcumin as an anticarcinogenic agent. One can imagine a scenario where molecules of this type act as an interphasic shuttle for radical repair.

Acknowledgments.

Experiments were performed at the Paterson Institute for Cancer Research Free Radical Research Facility, the Christie Hospital NHS Trust, Manchester. The Facility is funded under the European Commission TMR PROGRAMME - ACCESS TO LARGE SCALE

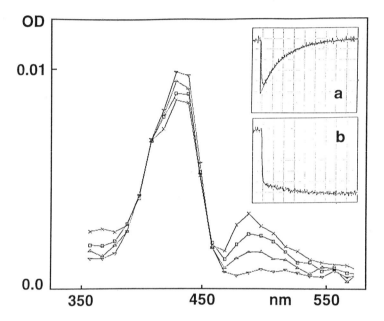

Figure 5. Transient absorption spectra measured 81 μs (X), 171 μs (□), 327 μs (Δ) and 576 μs (∇) after pulse radiolysis of a solution of curcumin (2.0×10^{-5} mol l^{-1}), sodium azide (2.0×10^{-2} mol l^{-1}) in N_2O-saturated pH 7 buffer containing Trolox (2.0×10^{-5} mol l^{-1}). Insets: time dependence of (a) transient decay at 490 nm with first-order fit, $k' = 3.3 \times 10^3$ s^{-1}, 0.4 % absorption/division, 150 μs/division and (b) corresponding growth of absorption at 430 nm with first-order fit, $k' = 3.2 \times 10^3$ s^{-1}, 0.4 % absorption/division, 150 μs/division.

FACILITIES, Grant ERBFMGECT 950084 - Access to PICR FRR Facility. This work was supported by NIH Grant 1 R15GM4435-01A1 (VSS) and the EPSRC (U.K.).

Literature Cited.

1. Tonnesen, H.; De Vries, H.; Karlson, J.; van Henegouwen, G. G. *J. Pharm. Sci.* **1987**, *76*, pp. 371-373.
2. Dahl, T. A.; McGowan, M. A.; Shand, M. A.; Srinivasan, V. S. *Arch. Microbiol.* **1989**, *151*, pp. 183-185.
3. Sharma, O. P. *Biochem. Pharmacol.* **1976**, *25*, pp. 1811-1812.
4. Nagabhushan, M.; Bhide, S. V. *Nutr. Cancer* **1986**, *8*, pp. 201-210.
5. Hartman, P. E.; Shankel, D. M. *Environ. Mol. Mutagen.* **1990**, *15*, pp. 145-182.
6. Nagabhushan, M.; Bhide, S. V. *J. Am. Coll. Nutr.* **1992**, *11*, pp. 192-198.
7. Gorman, A. A.; Hamblett, I.; Srinivasan, V. S.; Wood, P. D. *Photochem. Photobiol.* **1994**, *59*, pp. 389-398.
8. Hayon E.; Simic, M. *J. Amer. Chem. Soc.* **1970**, *92*, pp. 7486-7487.
9. Butler, J.; Land, E. J.; Swallow, A. J.; Prutz, W. *Radiat. Phys. Chem.* **1984**, *23*, pp. 265-270.
10. Gorman, A. A.; Hamblett, I. *Chem. Phys. Lett.* **1983**, *97*, pp. 422-426.

INDEXES

Author Index

Affiliation Index

Subject Index

Bestsellers from ACS Books

The ACS Style Guide: A Manual for Authors and Editors
Edited by Janet S. Dodd
264 pp; clothbound ISBN 0–8412–0917–0; paperback ISBN 0–8412–0943–X

Writing the Laboratory Notebook
By Howard M. Kanare
145 pp; clothbound ISBN 0–8412–0906–5; paperback ISBN 0–8412–0933–2

Career Transitions for Chemists
By Dorothy P. Rodmann, Donald D. Bly, Frederick H. Owens, and Anne-Claire Anderson
240 pp; clothbound ISBN 0–8412–3052–8; paperback ISBN 0–8412–3038–2

Chemical Activities (student and teacher editions)
By Christie L. Borgford and Lee R. Summerlin
330 pp; spiralbound ISBN 0–8412–1417–4; teacher edition, ISBN 0–8412–1416–6

Chemical Demonstrations: A Sourcebook for Teachers, Volumes 1 and 2, Second Edition
Volume 1 by Lee R. Summerlin and James L. Ealy, Jr.
198 pp; spiralbound ISBN 0–8412–1481–6
Volume 2 by Lee R. Summerlin, Christie L. Borgford, and Julie B. Ealy
234 pp; spiralbound ISBN 0–8412–1535–9

From Caveman to Chemist
By Hugh W. Salzberg
300 pp; clothbound ISBN 0–8412–1786–6; paperback ISBN 0–8412–1787–4

The Internet: A Guide for Chemists
Edited by Steven M. Bachrach
360 pp; clothbound ISBN 0–8412–3223–7; paperback ISBN 0–8412–3224–5

Laboratory Waste Management: A Guidebook
ACS Task Force on Laboratory Waste Management
250 pp; clothbound ISBN 0–8412–2735–7; paperback ISBN 0–8412–2849–3

Reagent Chemicals, Eighth Edition
700 pp; clothbound ISBN 0–8412–2502–8

Good Laboratory Practice Standards: Applications for Field and Laboratory Studies
Edited by Willa Y. Garner, Maureen S. Barge, and James P. Ussary
571 pp; clothbound ISBN 0–8412–2192–8

For further information contact:

American Chemical Society
1155 Sixteenth Street, NW ✦ Washington, DC 20036
Telephone 800–227–9919 ✦ 202–776–8100 (outside U.S.)

The ACS Publications Catalog is available on the Internet at
http://pubs.acs.org/books